FIELD GUIDE TO THE
ANIMALS
OF BRITAIN

ANIMALS OF BRITAIN

was edited and designed by
The Reader's Digest Association Limited, London.

First edition Copyright © 1984
The Reader's Digest Association Limited,
11 Westferry Circus, Canary Wharf, London E14 4HE
www.readersdigest.co.uk

Reprinted with amendments 2001

Copyright © 1984 Reader's Digest Association Far East Limited
Philippines Copyright 1984 Reader's Digest Association Far East Limited

Printed in Italy

ISBN 0 276 42503 0

The picture of the badger family on the cover was painted by Peter Barrett

READER'S DIGEST
NATURE LOVER'S LIBRARY

FIELD GUIDE TO THE
ANIMALS
OF BRITAIN

PUBLISHED BY THE READER'S DIGEST ASSOCIATION LIMITED

LONDON · MONTREAL · NEW YORK · SYDNEY

Contributors

The publishers wish to express their gratitude to the following people for their major contributions to this Field Guide to the Animals of Britain

PRINCIPAL CONSULTANT
Dr Pat Morris

OTHER CONSULTANTS
Lawrence Alderson M.A. (Agric.)
Dr Trevor Beebee
Norma Chapman B.Sc., M.I. Biol.
Dr Stephen Harris

CARTOGRAPHY
The distribution maps of animals are
based on information supplied by
The Biological Records Centre of
The Institute of Terrestrial Ecology,
and were prepared by
Clyde Surveys Limited, Maidenhead

ARTISTS

Peter Barrett	Rosalind Hewitt
Dick Bonson	H. Jacob
Jim Channell	Robert Morton
Kevin Dean	David Nockels
Brian Delf	Eric Robson
Sarah Fox–Davies	Jim Russell
John Francis	Gill Tomblin
Tim Hayward	Libby Turner B.A. Hons.
	Phil Weare

A full list of the paintings contributed
by each artist appears on page 320

PHOTOGRAPHERS
For a full list of acknowledgments to
the photographers whose work appears
in this book, see page 320

Contents

Understanding animals

The Animal Kingdom consists of all living things that are not plants. Although birds, fish and insects are therefore technically animals, in common usage the term is not generally applied to them.

This book covers all vertebrate (backboned) animals that live on land in Britain. It therefore includes all four-legged animals and also snakes and slow-worms (which are limbless vertebrates) as well as seals, which spend part of their lives on land and part in water. It does not include vertebrates such as fish and whales, which live entirely in water, nor does it include invertebrate land animals – creatures such as slugs, snails, worms and woodlice, which have no bony skeleton. These are dealt with in companion volumes.

The land animals described in this book belong to one or other of three major groups: mammals, amphibians or reptiles, which have each evolved different ways of using energy and reproducing their kind.

Mammals (which include man) have hairy or furry bodies and feed their newborn young on milk. They are described as warm-blooded because their body temperature is normally higher than that of their surroundings – usually within a range of 90–104°F (32–40°C).

Because they are able to maintain their own body heat, mammals can be active at any time. Muscle, nerve and digestive action increases at higher temperatures, and animals must have a certain body temperature before the system can be fully active.

Amphibians are animals such as frogs, which have soft, moist skins and can breathe air on land or absorb oxygen through the skin when under water. They live in damp places on land, in spring taking wholly to the water for a period in order to breed. They lay eggs that hatch into tadpoles, which live and grow entirely in the water for the first few weeks of life until they are transformed into tiny adults ready for life on land.

Reptiles are animals such as snakes, which have dry, scaly skins. They have lungs and breathe air, like mammals, but do not produce milk to feed their young. Some lay eggs and others give birth to live young.

Both amphibians and reptiles are 'cold-blooded' – that is, their body temperature varies according to that of their surroundings. They are fully active only when outside warmth raises their body temperature to a high enough level, which is generally about 77–90°F (25–32°C).

Plant-eating animals are known as herbivores. Some, such as sheep, are totally herbivorous; others will occasionally take different food – grey squirrels, for instance, will take birds' eggs. Animals that feed almost entirely on flesh are described as carnivores, and animals that will feed on plants or flesh as the opportunity offers are called omnivores. Although animals that eat insects are called insectivores, the term can be confusing because all insect-eating animals, bats for example, do not belong to the classification order Insectivora.

Animal classification

The smallest natural group to which an animal belongs is termed the species. Animals of the same species are able to interbreed and produce fertile offspring. Those of different species do not normally do so.

Although an animal's common name usually indicates its species – weasel for example – names can vary in different parts of the country, so animals are given a scientific name based on Latin. The weasel's scientific name is *Mustela nivalis*, of which *nivalis* is the species name and *Mustela* the genus name – a name shared by other closely related animals of a different species. So the related stoat is *Mustela erminea*.

When more than one genus closely resembles another, they are grouped together to form a family. The stoat and weasel belong to the Family Mustelidae. Various families are also grouped into related orders. The mustelids all belong to the Order Carnivora – flesh-eating animals.

It is usual for scientific names to be printed in italic type. The genus name is always given a capital letter, the species name a small letter.

With domestic animals such as the cow, there may be many different breeds within a single species; these breeds have been developed by man for particular purposes. A Jersey cow, for example, is a dairy breed developed on Jersey, and looks quite different from a Hereford, developed as a beef breed. Both belong to the same species and have the scientific name *Bos* (domestic).

The animals of Britain

Of the world's 5,000 species of mammals, only about 60 wild species live in mainland Britain – even fewer in Ireland. The exact number of wild species is difficult to establish because some are rare and nocturnal and hard to identify, especially bats.

The six species of common domesticated mammals that are farm or working animals – cattle, sheep, goats, pigs, horses and dogs – are included in this book. Farm animals also include poultry, and a number of fowl, turkeys, ducks and geese are described – the only animals in the book that are not mammals, reptiles or amphibians. Ponies, some of which are half-wild and some domesticated, are included in both the wild and farm sections.

Throughout the world there are more than 3,000 species of amphibians. Britain has 12 recorded species of frogs, toads and newts. Of these, five are present in very small numbers in a few colonies following deliberate or accidental introductions; they are the edible and pool frogs (very similar to the marsh frog), the tree frog, the midwife toad and the South African clawed toad.

There are only six species of British reptiles – three snakes and three lizards – out of a worldwide total of more than 3,000. Only one British species – the adder – is poisonous.

How to use this book

The main part of this book features 69 wild land animals of Britain, species by species. There are 22 different breeds of domestic animals, with many more described in less detail. The wild species are divided into three major groups – mammals, amphibians and reptiles, but the various domestic animals are all grouped together.

The mammals are further grouped mainly according to their classification Orders. An identification key appears on pp. 26–31, and this will direct you to the pages for the group.

Recognition characteristics are given for each species, and also their habits, life-span and a guide to habitat – the type of country in which the animal normally lives. As some groups of animals are very similar in appearance, there are a number of look-alike charts showing the differences to look for.

Recognition characteristics include an indication of the animal's size; where the male and female are appreciably different in size, this is stated. Individual animals vary in size, so the sizes are a guide only.

For smaller animals such as mice, size is indicated by head and body length, with tail length separate. For larger wild animals, such as deer, size is indicated by shoulder height. For farm animals such as cattle, size (where given) is indicated by weight, because many breeds are very similar in shoulder height but appreciably different in bulk.

The length of life given for any animal is only an indication of how long it might be expected to live in reasonable conditions. It is not a true average life-span for the species. Many wild animals die not long after birth, but the survivors may live for several years.

As many animals are active only after dark, tell-tale signs of their presence are more often seen than the animals themselves. The fieldcraft section (pp. 228–39) gives some idea of the signs to look for in various types of country. Throughout the book there are also a number of features showing how the life-style of certain animals is related to their surroundings.

To help you understand how an animal's observed behaviour is linked to its body structure and life-style, pp. 8–25 outline some reasons behind the different ways in which animals feed, breed and behave.

Using the maps

Distribution maps are given for each of the wild species featured in the book, and are based on the best and most up-to-date surveys. On each map, the area coloured in green is the area of country where the species can be reasonably expected to be found within its preferred habitat, which is indicated in the caption below.

For example, the brown hare will be found mainly in open fields in the green area. The grey squirrel is likely to be absent from quite large stretches of moorland within the green area on the map, but they are too small to be indicated.

Why skin is furry, scaly or bare

An animal's skin is adapted to its needs and its way of life. Typically, mammals have an insulating covering of fur – fine, closely packed hairs – because their body heat is generated internally and they need to prevent its loss. They need a thick coat to keep them warm in winter and a thinner coat in summer, so they moult to change the coat once or twice a year. Moulting is triggered by seasonal changes in day length and temperature, and controlled by hormones in the blood.

Reptiles and amphibians such as snakes and frogs absorb heat from their surroundings. A furry skin would impede the process, so reptiles have thin skin protected by thin scales, and amphibians have bare skin.

Mammals such as the mink and the otter, which spend a lot of time in the water, are in particular danger of losing body heat. To avoid this they have a furry coat which is effectively in two layers. The under layer consists of very fine hairs. These trap air against the skin, providing insulation and keeping the animal warm. On top is a layer of longer hairs that provide a sleek and stream-lined shape, and keep the underfur dry. It is the fine underfur that makes the pelts of these aquatic animals so valuable for high quality garments, and was the reason for the introduction of mink, musk rats and coypu into Britain.

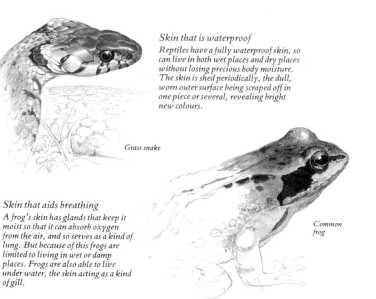

Skin that is waterproof
Reptiles have a fully waterproof skin, so can live in both wet places and dry places without losing precious body moisture. The skin is shed periodically, the dull, worn outer surface being scraped off in one piece or several, revealing bright new colours.

Grass snake

Common frog

Skin that aids breathing
A frog's skin has glands that keep it moist so that it can absorb oxygen from the air, and so serves as a kind of lung. But because of this frogs are limited to living in wet or damp places. Frogs are also able to live under water, the skin acting as a kind of gill.

Spiky hairs for protection
A hedgehog's spines are hairs modified to perform a protective role. They can be made to point in any direction, forming a protective jacket like barbed wire that few predators will breach. Spines cannot be moulted all at once or they lose their value. They are lost and re-grown one at a time over long periods.

Fat layers for warmth

Seals live in the water but do not have a double-layered coat like mink or otters. They are kept warm by thick layers of fat beneath the skin. Insulating underfur might spoil their streamlined shape, and when diving deep its effect would be reduced because the air would be squeezed out. Seals have short, stiff guard hairs that protect the skin from abrasion by rocks and sand.

Grey seal

A coat that changes colour

Stoats and mountain hares change their coats in winter, and many go white at the same time; this is possibly triggered by sudden cold. The advantages of the change of colour may be camouflage in snowy surroundings and also better heat conservation, as white radiates less.

Stoat

Common shrew

A new coat twice a year

Shrews moult twice a year. The short summer coat is shed in October, starting at the rump and moving forwards. It is replaced by longer winter fur. This is moulted in spring, starting at the head, and replaced by shorter fur for the warmer months. Many rodents also moult twice a year.

A new coat each summer

Foxes moult once a year, in early summer. Their thick winter coat is replaced by a shorter one that makes them look slimmer and longer-legged. As cold weather begins in autumn, more hairs grow and thicken the coat. As the density increases the hairs cannot all lay flat, and stand out from the skin, giving the fox the robust, stocky appearance shown here.

Specially developed underfur

Domestic sheep have been specially bred to develop their exaggerated curly underfur, which now forms their whole coat. It is this 'wool' that is spun into yarn or pressed into felt. Long, straight guard hairs are not wanted, as they cannot be linked up to form threads or felt, and would weaken the material.

9

How all land animals are built

Although many of Britain's land animals – mammals, reptiles and amphibians – look very different from each other, they have the same basic bone structure, even creatures as different as a frog and a ferret. All have evolved through millions of years from the same ancestors, so are variations on the same theme. The standard pattern can be seen in the rabbit skeleton shown under *How to identify bones* (p. 238), except that British snakes and slow-worms have no vestiges of leg bones. In all four-limbed animals, the forelimb joints compare with those of the human arm, and the hind-limb joints with those of the human leg. The similarities in forelimb bones for a variety of animals that differ considerably in size and shape are shown below.

Just as land animals have similar skeletons, so too are they similar in the layout of their muscles, in their digestive systems, and in their arrangements for breathing. But mammals have more highly developed and efficient blood-circulation systems than reptiles and amphibians, which gives them the ability to be much more active.

Common lizard

Typical foreleg bones

The bones in a badger's foreleg are typical of many land animals. It has a large shoulder blade, a single upper bone and twin bones in the lower leg. Small bones form the wrist, and each of the five toes has four bones, the end one with a horny claw attached.

Legs that spread sideways

A lizard's foreleg has the same bone pattern as a badger's, but its legs are spread sideways rather than held downwards. Newts are like lizards in posture but do not need such robust bones as they spend a lot of time buoyed up in water.

Forelegs with extra power

A digging animal such as a mole needs powerful muscles, which are anchored to prominent projections on bones. The relative sizes of bones are adapted to give extra leverage. A mole's upper foreleg bone is enormous, and it has a front foot that looks rather like a huge hand but is not very flexible.

Forelimbs that are flippers

Most of the forelimb of a seal is buried in its body, but it still has the standard bone pattern. The five toes, each with a long claw, are joined closely to form a flipper. This is used to steer in the water, but not for walking on land.

Common seal

Forelimbs that are wings

In a bat's forelimb, the four digits, or fingers, are enormously extended to carry the wing membrane between them. The fifth digit, or thumb, is not linked to the membrane and has a large claw used by the bat to hang or pull itself along.

Daubenton's bat

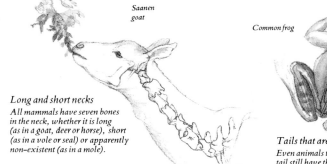

Saanen
goat

Common frog

Long and short necks

All mammals have seven bones in the neck, whether it is long (as in a goat, deer or horse), short (as in a vole or seal) or apparently non–existent (as in a mole).

Tails that are not obvious

Even animals that apparently have no tail still have the bones that would otherwise form it. In frogs and toads the bones are joined as a single rod that lies within the pelvis. Snakes have quite short tails – generally about a third or less of their length.

Feet with two toes

Sheep – like cows, pigs, goats and deer – have two toes on each foot ending in a cloven hoof. The remnants of two more toes appear as dew claws higher up the leg. There is no trace of a big toe. As with horses, the two longest bones of the leg are fused to give extra strength.

Feet with one toe

A horse has only one toe on each foot, corresponding to the second finger of a human hand; each toe ends in a large, protective horny hoof. The two longest bones of the leg are fused to give the limb extra strength. Generally, the fewer the toes and the longer the leg, the faster an animal can move.

Behaviour and body structure

An animal's normal behaviour is linked to its body structure. Some animals have evolved a specialised body structure that is suitable for a particular way of life but limits what they can do. Some (known as fossorial) are built for digging burrows – moles, for instance. Moles are very efficient diggers but cannot climb and are poor swimmers. Squirrels are arboreal (built for living in trees); they climb well but cannot dig burrows and also swim poorly. Animals built to run fast (cursorial), such as deer, cannot climb trees or use their limbs with the adroit manipulation of a squirrel.

Some animals, however, are 'Jacks of all trades', having evolved a less specialised structure that allows them to do all sorts of things moderately well. Rats are a good example. They can climb, burrow, swim and run fast, although in none of these activities are they as efficient as the specialists. Yet they are adaptable and not tied to one particular environment or way of life. They may be seen anywhere – indoors, in trees, in burrows or swimming.

Swimming animals need to push hard with their feet to force themselves through the water. This is made easier for them by webbing between the toes, to provide a larger surface that can 'get a grip' on the water. In the water shrew this is achieved by a row of bristles on each of the hind toes, but fully aquatic animals such as the otter have all four feet webbed, and are powerful swimmers. Frogs have fully webbed hind feet, and are better swimmers than the more terrestrial toads, which have only partial webbing on the hind feet and none on the forefeet.

Grooming with the tongue
Cats comb their fur with the tongue because their claws, which are sharp and curved for seizing prey, are not suitable for combing. When not in use, the claws are drawn back over the toes so they do not become blunted. Deer and other hoofed mammals also groom themselves mainly by licking.

Limbs for fast movement
Horses, deer and most other hoofed mammals have developed long, strong limbs so they can run fast and jump well to escape predators. Their feet are strong but light, having only one or two toes, but cannot be used to hold food; the lower leg can be bent only one way, not twisted round.

Claws for grooming
Like many other mammals, a water vole uses the claws on its forefeet as a comb to groom its fur. Mammals that spend a lot of time in the water groom often because their bodies are less well protected against cold and water if their fur is matted.

Tails that flash a warning

A rabbit's bobbing white tail as it runs from danger gives a warning to other rabbits. Muntjac deer show the white underside of the tail when alarmed, and roe and sika deer fluff out their light-coloured rump hairs.

A tail as an extra limb

A harvest mouse uses the end of its tail to grip twigs and stems as it climbs among them. Its grip is strong enough to support its own weight, so the mouse can stretch to get at food otherwise out of reach.

Hind legs for springing away

Frogs, like rabbits, hares and wallabies, have large, powerful hind legs that help them to spring away fast from a standing start. As their front legs are much smaller, they cannot easily walk on all fours but move in characteristic hops and bounds.

Tails for brushing away flies

Horses and cattle can flick their long tails in all directions to brush away irritating flies. Their ability to scratch using their legs and hoofs is very limited.

Tails that aid balance

Although a black rat cannot grip with its tail like a harvest mouse, it loops and swings its tail to assist balance. This enables it to run fast up or down or along very narrow surfaces such as ropes. Brown rats and some mice use their tails in the same way.

13

Why animals have different kinds of teeth

All Britain's land animals have teeth, although those of amphibians such as frogs are very tiny. The teeth of mammals are divided into specialist groups – incisors for biting, which are at the front of the mouth, premolars and molars (or cheek teeth) for chewing and grinding, which are at the back, and in between canines for seizing prey.

This basic tooth pattern is often modified for a particular diet. For example, flesh-eaters (carnivores) need sharp canines for quickly seizing and stabbing prey, but plant-eaters (herbivores) usually have no canines because they do not need to seize prey. Plant-eaters tend to have broad teeth, ridged like files, so that they can shred and grind plants to extract all the nourishment.

Some basically carnivorous mammals, notably badgers, actually eat all sorts of plant food as well as flesh, and are called omnivores. Their teeth tend to be broader and better adapted for crushing and grinding than those of mammals that are more strictly carnivorous. Pigs are omnivores and have well-developed canines.

Carnivores, omnivores and insectivores have teeth in a continuous row along the jaw. Herbivores and rodents (gnawing animals such as mice) have a long gap between the incisors and cheek teeth.

Chewing the cud
Plant food is hard to digest, so sheep, goats, cattle and deer have a specially adapted, multi-chambered stomach called a rumen. Food is briefly chewed and then swallowed into the first chamber, where it is partly digested. Later, when the animal is resting, wads of food, or cuds, are returned to the mouth for further chewing. Then they are swallowed into a different part of the stomach for complete digestion. Sheep, like cattle, have only a pad at the front of the upper jaw. But their small, narrow muzzles are more delicate than a cow's and they can nibble short grass.

Teeth that grow constantly
Squirrels, like all rodents, have big incisors that grow constantly but are kept in trim by gnawing hard food such as nuts. Unless rodents gnaw frequently, their front teeth may grow too long and develop into a grotesque arc.

Red squirrel

How cattle eat grass
A cow uses its long tongue to wrench up a bundle of grass, so cannot easily feed on very short turf. It has incisors in the lower jaw only, and a tough pad formed of the upper lip in the upper jaw. Grass is very tough, and cattle, like all grazing and browsing animals, have hard, curved ridges on their molars for grinding it into shreds.

Teeth for chewing bones

Dogs, foxes and other carnivores turn their heads sideways and use their molars to bite off tough pieces of food or chew a bone. Compared with those of rodents, their incisors are small and weak.

Teeth for seizing prey

As with all flesh-eating animals, a fox has large canines between its incisors and molars. The canines enable it to seize and stab prey and inflict a serious or fatal wound before the victim can escape. Bats are carnivorous and have the same tooth pattern. They need to seize and stab active and relatively large insects.

Teeth that inject poison

An adder bites its prey with poison fangs. They are hollow teeth which act rather like hypodermic needles and inject poison from glands in the upper jaw into the victim. As snakes swallow their prey whole, they do not need teeth for chewing. They have tiny, sharp, backward-sloping teeth that prevent their prey escaping.

Plant-eaters that feed on leaves

Deer are natural browsers – they feed on the leaves and shoots of trees – but may also graze, especially in parks when they have eaten all the leaves they can reach. As with sheep and cows, they have a tough pad at the front of the upper jaw and incisors on the lower jaw only.

Grazers that can bite

Horses can crop grass much closer than cattle, and eat coarser vegetation. This is because they have incisors on both their upper and lower jaws, so can bite their food instead of plucking it. Horses do not chew the cud, so plant material is not thoroughly broken down and may be recognisable in their droppings.

Teeth for eating small prey

Hedgehogs, moles and shrews have lower-jaw incisors that point forwards at an angle. They are effective for picking up small prey such as woodlice, which are crushed on the sharply pointed molars.

Adder

Eyes without eyelids

Snakes have no eyelids. A transparent disc of skin permanently protects the eye and is regularly sloughed with the rest of the skin. Adders have vertical rectangular pupils, grass snakes and smooth snakes circular pupils, but the reason for the difference is not known.

Grass snake

How animals look at the world

An animal's eyes do not see the world in exactly the same way as a human's eyes. Humans can perceive distinct colours, but animals do not normally see colour to the same extent. Animals that are active by day, such as squirrels, are more likely to see colours. Most land animals, however, are active mainly at night, when they see not in colour but in shades of light and dark. Their eyes are much more sensitive than human eyes and can work at very low light levels. On a very dark night, animals can see at least as well as humans do in bright moonlight but no animal can see in total darkness.

Many nocturnal animals – cats, for example – have a special reflective layer at the back of the eye that helps to increase its efficiency after dark. When looking directly at a light, such as headlights, their eyes reflect the light and shine in the dark. A cow's eyes also act in this way because it has evolved from creatures that were nocturnal millions of years ago.

Eyes that see by day or night

A cat's eyes, like those of all land animals, each have a central part, or pupil, that is always black and acts as a window, letting light pass into the eye. In the dark the pupil expands to let in as much light as possible. In the light it contracts to protect the sensitive region at the back of the eye from receiving too much light. A cat's pupils contract to vertical slits, but those of most animals are round when contracted.

Eyes that are small and weak

Despite the expression 'blind as a bat', bats' eyes see well enough to aid them in navigation. Even the mole's two tiny eyes are not blind, as is often believed; they can tell dark from light, and can also recognise movement.

Long-eared bat

Eyes with rectangular pupils

Sheep and goats have pupils that are rectangular and horizontal. No one knows for certain whether the animals gain any advantage from this. Toads also have pupils of similar shape when contracted.

Grey squirrel

Red deer

Eyes for judging leaps

Squirrels need to judge distances accurately when leaping among tree branches, so their eyes face more to the front than those of most other rodents. They also have short noses that do not get in the way when they look directly forwards.

Eyes of a different colour

The fox is unusual in having an amber, or golden–orange, iris (the part of the eye surrounding the pupil). Most cats have a green iris, but in many animals it is dark brown, giving an impression of all–black eyes and making it difficult to tell if the pupil is expanded or contracted.

Weasel

Eyes for stalking prey

Predators such as weasels need to focus on their prey, so their eyes face more to the front than animals with all–round vision. Each eye has a field of vision that considerably overlaps with that of the other eye; this gives three–dimensional vision and permits the accurate judgment of distances. But the head must be turned to scan an area.

Whiskers that show the way

Rats – like cats and many other animals that spend a lot of time moving about in the dark – have long, sensitive whiskers on the nose so they can feel their way where the light is too poor to see properly. Cats also have whiskers on their brows – some on their elbows too. Moles have sensitive hairs on the tail that enable them to feel their way backwards along their tunnels without bumping into anything.

Brown rat

Eyes for seeing all round

Animals that live or feed in the open, such as deer, sheep, horses and rabbits, need all–round vision so they can keep watch for danger even with their heads down while feeding. Their eyes are long–sighted and set at the side of the head so that they can scan about 300° – to the front, sides and most of the rear – without turning the head. But they see less detail than human eyes.

17

The secrets of scent

Most land animals have a much more sensitive and discriminating sense of smell than humans. Smell is the sense they employ most, using it to find food and water, follow trails and recognise each other. Many animals, mammals especially, produce particular scents to aid communication with each other; they are made in scent glands – often by the activity of bacteria – in various parts of the body, such as under the root of the tail. The scent secretions are frequently greasy so that they will last and not be washed away by the first rain.

Dogs secrete scent in their urine which reveals their sex and identity to other dogs that sniff it. The sniffer can also tell whether the scent is new or old. Many animals, carnivores especially, use scent to mark territory (a defended area). Badgers, which live in sociable 'clans', mark each other with scent so that clan members can be distinguished from outsiders. They can probably also recognise clan members individually by their scent. Some animals, weasels and grass snakes for instance, use scent as a defence. When they are attacked, they eject a foul-smelling scent that distracts their assailant, and may even cause distress.

Scent from the face
Most bats have scent glands on the face, but on the noctule bat they are particularly noticeable. The secretions are strong-smelling and greasy, and may help sex and roost identification. Probably the female also recognises her own baby in the dark by its scent.

Noctule bat

Scent from the feet
Dogs scratch at the ground to mark territory with scent from glands between the toes. Some deer – fallow for example – have scent glands in the cleft of each hind foot; their secretions probably convey messages to other deer.

Dalmatian

Scent from the chin
A rabbit rubs the large scent gland under its chin against saplings, posts, stones or the ground, to mark them as part of its territory. Cats also use their chin or forehead to rub their scent against familiar objects.

Scent from the tail
Otters are among the many animals that leave scent marks as claims to territory. They often scent their droppings with secretions from glands under the tail, or sometimes rub their haunches on stones or logs to smear scent on them.

Common newt

Finding water by smell
Newts and other amphibians that hibernate on land make their way to ponds each spring to breed. They seem able to find their way to the water by its smell.

Scenting the air

When a fox waves its bushy tail, it may be wafting out scent from the gland on its upper surface. A squirrel's tail may perform a similar role. Squirrels certainly twitch their tails frequently when agitated or displaying to a rival.

Hunting by scent

Like all snakes, the grass snake constantly flicks out its tongue to 'taste' the air or ground in front of it and find its way about. In this way it detects the scent of prey.

Red deer

Messages between mates

When male red deer or sheep stretch out their necks and pull back their lips, they are sampling the air to detect scents. This behaviour, called 'flehmen', is common in hoofed mammals, which do it most at breeding time to tell when a potential mate is in season.

Singling out the baby

Most mothers will suckle only their own young. Ewes, cow seals and other animals that breed among large groups recognise their own baby in a crowd of others by its individual smell.

The significance of hearing

Most animals need acute hearing in order to survive. They must constantly listen to be aware of danger or to detect prey, especially as many are active at night when vision is limited. Often, mammals can move each ear independently, enabling them to scan all round while keeping still and undetected.

Many animals can hear ultrasonic sounds – that is, sounds too high-pitched for human ears. Bats have the best-developed hearing of all, and for them it largely takes the place of vision. A bat with its ears experimentally plugged with cotton wool is almost helpless and will fly only with extreme reluctance. Yet blindfolded bats can fly and catch food with little difficulty. Snakes are deaf, having neither ears nor eardrums. They are sensitive to vibrations through the ground.

In the wild, mammals are silent most of the time because it is safer not to attract unwelcome attention to themselves. Most will, however, growl, squeak, hiss or scream if attacked or frightened. Some species have special calls to communicate with others of their kind – calling a mate, for example.

Long-eared bat

Horseshoe bats

Snarling a warning
Dogs, like many other mammals, snarl and grind their teeth to warn off other animals. Often they also use visual signals such as erect fur, bared teeth and a crouching posture ready for springing to the attack.

Hunting by echoes
Bats emit ultrasonic sounds and locate prey (or obstacles in their path) when the sounds bounce back. Horseshoe bats emit sound through the nostrils that is directed and focused by the fleshy flap round the nose. Ordinary bats such as the long-eared emit sound through clenched teeth. All bats have relatively large ears in order to catch the slightest echo.

Listening for danger
The black rat's ears are constantly twitching to detect the slightest sound, especially when it is feeding. It needs to be alert to danger, such as a prowling cat. Like mice, rabbits and other animals preyed on by carnivores, it has large ears.

Calling to keep in contact
Domestic animals such as cows, which are safe from predators, are not as silent as most wild animals. They often call loudly to straying young, and to keep in contact with the herd. Grey seals are very noisy on their breeding grounds; a pup's bleat helps its mother to single it out.

Listening for prey

Predators such as the pine marten tend to have forward-facing ears that can focus on the same area as their forward-facing eyes, so bringing two senses at once to inspect in detail the scene ahead.

Calling a mate

Like other male frogs, male marsh frogs croak loudly at breeding time to attract females. Large vocal sacs amplify the voice to carry over a wide area. Frogs have no ear flap, but there is a large, sensitive eardrum behind the eye.

Bellowing for mastery

Red deer stags bellow loudly at rutting time to intimidate rivals and win the notice of females. The stags that bellow longest and most often acquire the largest harems. Fallow bucks and sika stags behave in a similar way.

Common dormouse

Squeaks as tracking signals

Nestling mice and other rodents give out tiny squeaks too high-pitched for humans to hear. The squeaks help the mother to find young that stray from the nest and keep the family together in the days before their eyes open.

The talking foxes

Foxes commonly bark, yap or howl at night to communicate with each other. They have a dozen or more distinctive calls, each with its own meaning, and are especially noisy in winter, during the breeding season.

21

Breeding and the care of the young

In wild animals, breeding is linked to the seasons. This ensures that the young are produced at the most advantageous time of year, when food is plentiful. In some mammals (badgers, stoats, grey seals, roe deer), the egg does not begin to develop in the womb as soon as it is fertilised but starts after a delay of some weeks. This allows courting and mating to take place at a convenient time but ensures that the young are born when conditions are suitable.

Grey seals, for example, gather on shore in large herds for breeding each autumn, and mating takes place soon after the birth of pups from last year's mating. The young take only about 250 days to develop, so if the egg were implanted at once, pups would be born well before the next autumn gathering. But implantation is delayed roughly 100 days so that the pups are born at the most suitable time.

Breeding in farm animals is arranged to suit the farmer, who selects the mating time and the animals to be mated. In this way farmers can develop commercially useful features such as fast growth, higher milk production or more vigorous offspring.

Cross-breeding for vigour
A cross-bred domestic animal (from parents of different breeds) is usually more vigorous and adaptable than its parents. Farmers use selective cross-breeding to produce vigorous offspring that combine particular qualities of different breeds. A blue cross pig, for example, has some of the hardiness and mothering instinct of its British Saddleback mother and some of the lean meat yield of its Large White or Landrace father. So it is a good-quality pig hardy enough to be kept outdoors.

British Friesian cow

Young hatched from eggs
Frogs, toads and newts lay eggs covered in jelly (spawn) in water. Reptiles lay eggs enclosed in a shell or membrane. Some, the grass snake for example, leave them in a warm place. In others, such as the adder, the eggs hatch inside the female so that live, active young are born.

Selective breeding to produce milk
Dairy cows such as the British Friesian have been specially bred to produce unnaturally large amounts of milk after calving. Calves are taken away at birth, and the cow, provided she is well fed, will give milk for far longer than she would for her calf — perhaps for ten months. She will release all her milk readily in the absence of her calf, an unnatural trait. Young cows are called heifers until they have their first calf at about 18 months old.

Young reared on milk
The grey seal, like all mammals, suckles its young on milk, which provides nourishment in the form of fats, sugars and proteins, and also helps to combat infection. All mammals care for their young until they are old enough to fend for themselves.

British Saddleback

Blue cross piglets

Bank vole

Young born in a nest or den

Voles and other small mammals give birth to blind, naked young. The mother selects a sheltered place to build a nest in which the large litter is kept warm and protected for a number of weeks until the young are developed sufficiently to survive outdoors.

Care from the father

A male fox will bring food to the den to help feed the female and her cubs. This is unusual; the males of most mammals play no part in rearing the young.

Young born in the open

For larger mammals such as fallow deer, making a nest and rearing a large family is not feasible. The young are born in the open, usually only one each year, and have a full covering of fur or hair. They open their eyes at birth and are quickly able to walk and run.

Preparing a breeding nest

Breeding nests are made by the female, generally from plant material. A doe rabbit plucks fur from her own chest to make a warm lining for the nest chamber of her burrow. The fur comes out easily because of hormone changes during pregnancy.

23

Patterns of sleeping and feeding

All animals need to rest and sleep from time to time, and do so in as safe a place as possible, such as within a herd or hidden in a nest or burrow. Safety, temperature and the availability of food all play a part in deciding when an animal is active and when it rests.

Most land animals sleep or rest during the day and seek food under cover of darkness. Larger animals such as deer have less need to worry about predators so can sleep at any time. Although some animals, such as shrews, need to eat almost constantly to keep alive, others do not necessarily eat every day. Snakes, for example, can live for a week or so on one meal. With cold-blooded animals – snakes, lizards and amphibians – activity depends on body temperature. Snakes and lizards rely mostly on the heat of the sun so are out only in daylight, and in cold weather are inactive. Frogs and toads are able to move about in the cold to a greater extent than reptiles, and can be active at night.

For many animals the food supply dwindles drastically in winter, so they feed well in autumn to build up fat reserves. Some, dormice for example, then spend winter in hibernation – hidden in a safe place with all their body functions slowed down.

Red squirrels

Safety in daylight
Squirrels are among the few wild mammals that are active only by day. Their food is easy to find in daylight, and leaping from branch to branch is safer when they can see where they are going. In the tree tops they have little to fear from predators, so have less need to be nocturnal than mice, for example.

Wood mouse

Safety in darkness
Wood mice, like other small mammals, are vulnerable to predators, and venture out only when it is dark. Voles, however, which live among dense cover, can risk going out by day as well as by night.

Life with the tide
Seals are active by day and night and sleep at any time, on land or in the water. The state of the tide, not light and darkness, rules their lives. They feed at high tide when there are shallows where they can find crabs and flatfish. When the tide is low, they find convenient spots to come ashore and rest.

Common seals

Horseshoe bats

Sleeping through winter

Bats, like hedgehogs and dormice, hibernate in winter because their food supply is scarce in cold weather. By staying inactive and at a low body temperature they conserve energy and reduce their need for food. Bats hibernate most efficiently in places with an even temperature of 36–43°F (2–6°C). If the site they first choose gets too warm or too cold as winter progresses, they may wake up and move to a more suitable place. Horseshoe bats, especially, need places that are also damp.

Safety in numbers

Sheep prefer to live in flocks and show signs of distress if they find themselves alone. Like other herd animals, they probably developed this social behaviour as a defence against predators, finding greater safety in numbers, especially in open country. Predators such as foxes and mink live alone or in pairs.

Day and night activity

Shrews use up so much energy that they have to replenish it by eating frequently. They are active both day and night, and alternately rest and forage every few hours. To avoid predators they normally stay out of sight in burrows or under dense vegetation.

Common shrew

Food by night

Hedgehogs, although they have little to fear from predators, come out mostly at night. This is because they feed mainly on nocturnal creatures such as worms, slugs and beetles.

A regular routine below ground

To the mole in its dark tunnels there can be little difference between night and day. It has several spells of activity within a 24 hour period – perhaps three or four, with rests between. But activity often occurs at similar times in successive 24 hour periods, suggesting that moles are not totally insensitive to time rhythms.

Identifying animals at a glance

All the animals described in this book live on land or at the water's edge. Many are shy, nocturnal or rare – often all three – so may be glimpsed only briefly. To aid recognition, this identification key shows animals as they might appear at first glance, and they are grouped according to their characteristics, except for farm animals, which are all grouped together. The key gives an indication of size and appearance and, in some cases, where the animal might be seen, as well as where to find it in the book.

Size is crucial in identification – note how an animal compares in size with a familiar creature such as a cat. Where it is and what it is doing are also useful clues to identity. Many smaller animals are most often seen only when dead. This gives a good opportunity to study their colour, fur texture and other features missed in a brief glimpse.

Carnivores   Pages 34–73

All are flesh eaters, including fish-eating seals. Many also eat other foods such as fruit and insects.

Wild cat

Fox   Pages 34–39
Looks like a slim, prick-eared dog, but with a very bushy tail. Often seen by day in all sorts of country and in suburban areas.

Cats   Pages 40–43
Striped, bushy-tailed, tabby-like wild cats are found only in Scotland. Feral cats (domestic cats living wild) occur all over Britain.

Badger   Pages 44–49
Resembles a stockily built dog, with short legs and striped face. Active mainly at night in woods, meadows, moorland.

Otter   Pages 50–55
A long, slim, flat-headed animal that swims and dives well. Found in undisturbed places in or near water. Scarce, except in Scotland and Ireland.

Pine marten   Pages 58–59
Looks like a short-legged cat with a thick tail. An uncommon forest and moorland animal. Rarely seen by day.

Grey seal

Polecat   Pages 60–61

The size of a small, short-legged cat.
Found in Wales. Paler forms elsewhere
are probably polecat-ferrets.

Mink   Pages 62–63

A bushy-tailed waterside animal that
swims and dives well. Often seen by
day. Smaller than an otter.

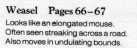

Stoat   Pages 64–65

Rat-sized, with a very long body that
undulates as it bounds. Often seen on
moors, farmland, roads and verges.

Weasel   Pages 66–67

Looks like an elongated mouse.
Often seen streaking across a road.
Also moves in undulating bounds.

Seals   Pages 68–73

Seen in coastal waters, rivers and
estuaries, or basking on rocks or
sand-banks. On land they move
awkwardly.

Hoofed mammals   Pages 74–107

Long-legged animals that feed on grass and leaves. All except horses
and ponies have cloven hoofs.

Red deer

Muntjac

Exmoor pony

Half-wild ponies   Pages 74–75

Unshod breeding herds inhabit
moors and forests, especially in
south and west of England and
Wales. All have owners. Colours
are very variable.

Deer   Pages 76–103

Long-legged animals that
range from muntjac (springer-
spaniel size) to pony-sized
red deer. Most males have
antlers, but not all year. Seen
wild or in parks.

27

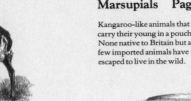

Marsupials   Pages 108–9

Kangaroo-like animals that carry their young in a pouch. None native to Britain but a few imported animals have escaped to live in the wild.

Feral goat   Pages 104–5
Looks like a shaggy sheep with widely spread horns. Found on cliffs, mountain-sides, in north and west.

Soay sheep   Pages 106–7
Brown with coiled horns. Found wild on some Hebridean islands. Some half-wild herds in England.

Red-necked wallaby   Pages 108–9
Moves in hops or bounds. When sitting upright is waist-high to a man.

Insectivores   Pages 110–31

Animals of varied appearance but very similar internal structure. All eat insects, but also worms, slugs and other small creatures.

Common shrew

Mole   Pages 110–15
Hamster size. Has large, heavily clawed forefeet.

Hedgehog   Pages 116–21
The size of a guinea pig. Has a spiny coat. Seen in farms, gardens, sometimes by day.

Shrews   Pages 124–31
Smaller than mice, and with long noses. Very active; squeak loudly.

Rodents   Pages 132–69

Gnawing animals, mainly active at night. Alike in having large front teeth that grow constantly.

Brown rat

Wood mouse

Field vole

Rats   Pages 132–5
Have thick, scaly tails; big eyes and ears. Usually in or near buildings. Brown rat common in hedgerows.

Mice   Pages 136–45
Small, with long, thin tails and big eyes and ears. Found indoors and outdoors.

Voles   Pages 146–51
Slightly bigger than mice. Chubby and round-faced with short tails, and ears buried in fur.

Common dormouse

Grey squirrel

Hamster

Dormice   Pages 160–3
Have fluffy tails and big eyes. Come out only at night, climbing among trees and bushes.

Squirrels   Pages 152–9
Bushy-tailed and rat-sized. Seen by day in or near trees. Grey species widespread, red mainly in north and west.

Alien rodents   Pages 164–5
Many pet hamsters and gerbils, and other foreign species from zoos, escape and establish wild colonies.

Hares and rabbits   Pages 166–77
Long-eared animals with very short tails. Their long hind feet give them a hopping or bounding gait. They are normally active by day or night.

Bats   Pages 178–201
Flying mammals with mouse-sized, furry bodies. At rest they hang head down in trees, lofts, caves.

Horseshoe bats   Pages 180–3
Have horseshoe-shaped skin round nostrils. Out only at night.

Lesser horseshoe bat

Brown hare

Natterer's bat

Hares   Pages 166–71
Look like large rabbits with big, staring eyes. Live above ground.

Rabbit   Pages 172–7
Usually found in colonies, mostly in lowlands. Often seen near the burrow in which it lives.

Ordinary bats   Pages 184–99
Have dog-like muzzles and dark or pinkish faces. Out only at night.

Amphibians and and reptiles   Pages 202–27

Amphibians live in damp places and lay their eggs in water. Reptiles are able to live in drier places and bear their young, or lay eggs, on land.

Common toad

Common newt

Common frog

Frogs   Pages 202–5
Amphibians with smooth, moist skin and long hind legs with feet fully webbed. Move in leaps.

Toads   Pages 206–9
Amphibians with rough, warty skin, drier than a frog's. Hind feet only partly webbed. Normally crawl.

Newts   Pages 210–11
Lizard-like amphibians with smooth skin. Tail flattened at sides. Walk slowly with wriggling movement.

Adder

Common lizard

Snakes   Pages 214–19
Limbless reptiles with dry, scaly skin and no eyelids. Move with a gliding motion from side to side.

Lizards   Pages 220–3
Small reptiles with dry, scaly skin. Tail rounded. Move with swift, darting wriggle.

Slow-worm   Pages 224–5
A snake-like legless lizard with smooth, dry, shiny skin. Unlike snake, has eyelids. Often found in compost heaps.

Farm animals   Pages 240–89

The animals most often seen are those kept and used by man. Many are intensively cared for and are never seen far from houses or farms. Others, hill sheep for example, roam almost as free as wild animals.

Guernsey cow
Suffolk sheep

Cattle   Pages 240–51
Colours and build help to distinguish breeds. Beef breeds are stocky, dairy breeds slighter and more angular.

Sheep   Pages 252–65
Breeds distinguished by size, build, wool length, face and leg markings, and type of horns (or lack of them).

British Saddleback pig

Saanen goat

Old English Game fowl

Embden goose

Pigs   Pages 266-9
Colour, ears (pricked or lop) and build help to distinguish breeds. Some are not hardy enough to live outdoors.

Goats   Pages 270-3
Different breeds are distinguished by colouring and coat pattern. Some may also be bearded and horned.

Poultry   Pages 274-7
Some hens, ducks, geese, seen outdoors. Colouring and build distinguish different breeds.

Shire horse

Dales pony

Border collie

Dogs   Pages 278-83
Size and appearance more varied than in any other species. Some working breeds seen with livestock.

Horses and ponies   Pages 284-9
A few ponies and heavy horses may be seen at work on farms. Size, build and colouring help to distinguish breeds. Heavy horses mostly seen at agricultural shows or pulling brewery drays in towns.

31

ANIMALS
OF BRITAIN

Surplus food is often buried for later, but another fox in the family group may find it first. The fox uses its forefeet to dig a hole, its nose to push soil over the cache.

Prey such as voles may be detected by sound. The fox leaps on the spot the sound came from, pinning the prey with its forepaws.

About May foxes begin to moult their winter coats, and scratch to remove loose fur. The moult spreads slowly along the back and hindquarters and they look rather untidy for most of the summer.

Foxes can climb well. They will use the lower branches of a tree as a daytime resting place that also serves as a good vantage point.

Although foxes from the same group forage alone, they may meet during the night for play or mutual grooming. They greet each other with wagging tails and whickering noises.

34

Amber eyes

Bushy tail

Foxes are very adaptable. Some live in city suburbs, and may occasionally glean scraps from dustbins if other food is not readily available.

Fox *Vulpes vulpes*

Many tales are told of the cunning of the fox in eluding the hunt or catching prey. Foxes are indeed resourceful animals, and manage to thrive in all sorts of places. They are mostly active at night, when they forage for whatever food is available, scavenging from carcasses or killing small mammals, especially field voles and rabbits. In summer they catch large numbers of beetles, and in autumn feed on fruit. Foxes in coastal areas forage along the shore for crabs and dead fish or seabirds.

Alert and wary, foxes have acute hearing and a keen sense of smell. Their eyes are quick to detect movement but do not see stationary objects so well. A fox looks its best from October to January when its coat is full and thick. For most of the summer it undergoes a protracted moult. Although foxes are normally seen alone, they live in family groups usually made up of a dog fox (male), a breeding vixen (female) and her cubs, and perhaps one or two non-breeding vixens from previous litters. A den or earth is used at breeding time. A den may be in a rock crevice or under tree roots. The vixen may dig her own earth or enlarge an abandoned burrow. At other times, foxes usually shelter above ground. Few live more than eight years.

Widespread and abundant in many habitats, ranging from city streets to mountains.

The fox's amber eyes and bushy tail are distinctive. Its coat may vary from yellow-brown to red-brown, and its rump is often silvery with white-tipped hairs. The lower legs and the backs of the ears are black, the tail-tip often white. Dog 26 in. (66 cm) head and body, 15 in. (38 cm) tail. Vixen smaller.

Rearing a family of fox cubs

Foxes breed only once a year. The mating season lasts from Christmas until about February, when courting foxes may be heard emitting short triple barks, or may shatter the silence of the night with unearthly screams as a vixen calls to a potential mate. The dog fox and vixen hunt and travel together for about three weeks. Towards the end of this period they may mate several times. The vixen is pregnant for about 53 days, the peak period for births being around mid-March.

A litter of cubs is born on the bare soil of the den or earth; the vixen makes no nest. The cubs open their eyes when 10-14 days old, and take their first solid food – often regurgitated by their mother – when they are from three to four weeks old. A week later they emerge from the den for the first time, and their dark brown cub coats start to change colour. By about eight weeks the coat is red-brown. Non-breeding vixens may help to rear the cubs. Throughout summer the cubs stay together as a family, reaching adult size about September. Young vixens may stay with the family group, but young dog foxes leave in autumn or winter to find their own territories. At this time many young foxes are killed on the road by cars.

During the mating season the dog fox, tail held straight out, will follow the vixen for long periods. They may be seen during the daytime.

Four-week-old cub

Cubs grow rapidly. At four or five weeks old their blue eyes slowly change to amber and their coats begin to go reddish-brown.

Eight-week-old cub

A litter of four or five round-faced, short-eared cubs is born in March or April. They are born blind and are covered with fur of a deep chocolate brown.

If disturbed, the vixen moves the cubs to another earth. Cubs up to six weeks old are carried in her mouth, one at a time, each held by the scruff of the neck.

At the end of the year, some of the cubs leave to find their own territories. An adult will sometimes drive a young fox away from the group.

The vixen stays in the den or earth with her cubs until they are two weeks old, the dog fox bringing her food. After that she spends more time outside.

As the cubs grow up, they fight and squabble more and more often. Sparring cubs will stand on their hind legs and push each other.

The fox as a town-dweller

There are foxes living in most towns and cities from Nottingham southwards. They are commonest in areas with houses built between the two World Wars. In northern England and Wales urban foxes are much scarcer. There are some in Scotland – in Edinburgh and Glasgow – and also in Ireland – in Belfast and some cities of the Republic.

Town foxes are sometimes about in the daytime, not at all worried by the presence of man. But usually they rest by day, perhaps under a garden shed or in a sunny spot on an easily accessible roof. At night they forage over a small area, probably less than $\frac{1}{4}$ sq. mile (0·6 sq. km) in extent, mostly looking for scraps, windfalls or worms, but occasionally taking a pet rabbit or guinea pig. A vixen chooses her breeding den in late winter. Under a garden shed is a favourite spot, but she may use a pile of rubbish or dig an earth. Cubs may be seen playing outside from late April or early May. Some will readily take food from the hand. A pile of bones, chip papers and other food remains often accumulates outside the den in May, but from late June the cubs nest above ground during hot weather, behind bushes or crates or in similar spots, and the den is gradually abandoned.

A fox may rob a bird table of crusts or bacon rind if it is easy to climb or leap on. Worms often surface on a lawn on a warm, damp night, providing a feast for a foraging fox.

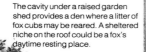

The cavity under a raised garden shed provides a den where a litter of fox cubs may be reared. A sheltered niche on the roof could be a fox's daytime resting place.

Urban foxes and domestic cats rarely interfere with each other's activities. If there is a confrontation, the fox is quite likely to be the first to back down.

Foxes are scavengers and do raid dustbins; but only occasionally – not often, as is generally believed.

A suburban back garden after dark may be a hunting ground for one or two foxes, or may be their permanent home. They prefer a garden that is not too tidy, offering plenty of shrubs or clutter for shelter.

Before mating, the male will follow the female and nuzzle her flanks. In Britain wild cats mate in the first two weeks in March. The female is pregnant for 65 days.

There are usually four kittens in a litter, born in May. They emerge from the den at about four or five weeks old, and often play together. They start learning to hunt with their mother when about nine weeks old and are weaned at about four months old.

A wild cat usually rests during the day. It may spend a sunny day basking on a tree branch or rocky outcrop, where it can keep a look-out over its surroundings.

Ferocious when cornered, the wild cat faces an adversary with back hunched, ears flattened and teeth bared, spitting its defiance. It will fight off a small dog.

The den is sited where there is a good view of the area around. It may be in a rock pile, in an old fox earth or under a tree stump.

When a wild cat scratches a tree it may not be just to sharpen its claws or stretch. It may also be marking its territory with scent secreted by its foot glands.

Wild cats put on a layer of fat in autumn in order to survive the harsh Highland winter. When food is scarce they may wander beyond their usual range.

Rounded tail-tip

Dark body stripes

Although it resembles a domestic tabby, the wild cat is slightly larger with longer, softer fur. It has black or grey body stripes and a bushy tail with a blunt, rounded tip. The tail has from three to five dark rings. Male up to 36 in. (90 cm) head and body, 12 in. (30 cm) tail. Female slightly smaller.

Domestic cat

Wild cat

A wild cat's tail has a rounded tip, a domestic cat's is pointed. A wild cat has complete body stripes, but a tabby is blotched. Because wild and feral cats interbreed, not all wild cats are so distinctively marked.

Found in Highlands. Spread southwards masked by interbreeding with feral cats.

Wild cat *Felis silvestris*

Remote Highland forests and moors are the home of the wild cat, which was once found throughout Britain, although not in Ireland. It was persecuted as vermin and almost died out in the 19th century, surviving only in isolated areas north of Scotland's Great Glen. Since the 1920s it has been slowly spreading southwards again. The increase in Scotland's coniferous forestry plantations may have aided its recovery.

Wild cats are most active at dawn and dusk, hunting alone or in pairs. Rabbits, mountain hares and small rodents and birds are their main food. The cats either lie in ambush to pounce on their prey, or stalk it until quite close and then attack with a final rush. The male defends a territory which, on average, covers about $\frac{1}{4}$ sq. mile (0·6 sq. km). Wild cats usually keep to one mate, but spend much of their time alone. Normally they breed only once a year. Second litters born in late summer are probably the offspring of hybrids between wild and feral cats. As feral cats are generally smaller, such interbreeding may account for the small reduction in the size of wild cats during this century. Young wild cats are independent at about five months and fully grown by about ten months. The life-span is up to 12 years.

Many farmyard cats are feral, living free of human control and hunting for their own food.

Members of a cat colony rub against one another to transfer scent from a gland on the top of the head. This helps them to recognise one another.

Urban feral cats will hunt small birds and mammals to add to their diet of scraps provided by humans.

In central London some feral cats live in basement yards and feed on scraps put out by residents.

A threatening pose or a blow of the paw usually drives off an intruder. Serious fights are rare.

Prominent places in the territory are scent-marked by rubbing with a gland under the chin, or by spraying urine.

A female bears her kittens in some warm, quiet nook within the colony's territory, perhaps in a hayloft or even in an old coat among the cars in a breaker's yard.

Pointed tail

Tabby blotching

May be present in any part of Britain. Usually most numerous in urban areas.

There are free-living domestic cats in both town and country. Most wild-born animals are pure black, black with white markings or tabby. Feral tabbies are more blotched than true wild cats and have pointed tails. Size varies greatly. Average about 20 in. (50 cm) head and body, 12 in. (30 cm) tail.

Feral cat *Felis* (domestic)

Feral cats are domestic cats that have reverted to living wild. Some are lost or abandoned pets; others are descendants of such pets and have lived wild all their lives. Although feral cats may be solitary like their domestic counterparts, they usually live in colonies composed of related animals. In rural areas there are small colonies of feral cats living in farm outbuildings, but most live in towns – not in residential areas where pet cats are numerous, but in the grounds of institutions such as hospitals, as well as in factories, dockyards and even city squares. An average colony has about 15 cats, but some are much larger; the Portsmouth Dockyard colony sometimes numbered 300.

A feral cat colony establishes an order of rank among its members and claims its own territory, driving out other cats. Females outnumber males by about three to two, and each year a female may bear three or four litters of about three kittens. Many kittens die of cat 'flu and other diseases, or in accidents. Rural cats feed mainly on small mammals and birds and must be successful hunters to survive. Cats in built-up areas feed mostly by scavenging, depending on household and restaurant waste or food put out for them by cat lovers.

43

Badgers keep to the same well-beaten paths through their territory. Wiry black-and-white hairs on a fence indicate a badger route. To prevent badgers tearing up rabbit-proof fencing, forestry fences may have special badger gates with heavy flaps.

Straw, dry leaves, bracken or green plants are used as bedding. The badger gathers them between forepaws and chin and shuffles backwards to the sett entrance.

Badgers use trees near the sett to sharpen their claws and clean mud off their paws, particularly on heavy soils. They prefer rough-barked trees such as elders and oaks.

In some urban areas, badgers will enter gardens to feed on household scraps, crops, windfall fruit or dustbin refuse, and may do damage.

Badgers belonging to the same group scent each other to aid recognition, a process known as musking. A musking badger backs on to another with its tail raised to secrete an odorous liquid from a gland under the tail.

White-tipped ears

Striped head

Strong forepaws

Badgers like to live in undisturbed woodland that provides well-drained, easily dug soil, plenty of undergrowth for cover and a good supply of food.

Badger *Meles meles*

There are badgers in most parts of the country, and some places have been named after them, such as Brockhall in Northamptonshire – *brocc* being Old English for badger. Some inhabit urban areas, notably on the south coast and in Essex, London, Bath and Bristol. Generally they are active at night and rarely seen. Badgers live in extensive burrow systems, or setts, dug out with their broad, powerful forepaws. Setts are usually in woodland, sometimes in fields or rubbish dumps, and include sleeping chambers where there is regularly changed bedding. Each sett is occupied by a group of one or two families. The group forages within an established territory, defended against outsiders, which has well-defined paths between the sett, feeding grounds and latrines – dung pits dug singly or in groups.

Earthworms are the badger's main food, supplemented by cereals, beetles, fruit in autumn, and some mammals, particularly young rabbits dug out from their burrows. Badgers will also dig out and eat the contents of wasp and bee nests. A few badgers probably survive for up to 15 years. Unauthorised killing is illegal, but control may be necessary, for example, where serious property damage results from the badger's burrowing activities.

Commonest in south and west. Scarce in East Anglia, parts of Scotland, urban Midlands.

The badger emerges cautiously from its burrow (or sett), sniffing for danger, soon after dusk. Its black-and-white striped head with small, white-tipped ears, is distinctive. Strong forepaws with long claws make it a powerful digger. Male 30 in. (76 cm) head and body; 6 in. (15 cm) tail. Female smaller.

45

In parts of the south-west the badger population is high, and setts are often close to cattle pastures. Some badgers are infected with bacteria that cause tuberculosis in cows, and may be infecting cattle. The incidence of tuberculosis is five times higher in the south-west than elsewhere, although less than one cow in a thousand is affected.

Badgers, cattle and tuberculosis

In 1971 it was discovered that a dead badger in Gloucestershire was infected with the bacteria that cause tuberculosis in cattle. Since then, infected badgers have been found in various parts of south-west England and occasionally in other places. The lungs and kidneys are the organs most prone to infection, so the infected badger may spread the bacteria through its breath, droppings and urine. Tuberculosis often spreads rapidly among badgers. In a few areas, up to 20 per cent of the badger population may have tuberculosis. Infected animals can survive for a long time and spread the disease widely.

Infected cattle suffer most from lung, not kidney, tuberculosis. They probably catch it by sniffing at foraging badgers or at pasture contaminated by them. Despite stringent official efforts to eliminate tuberculosis in cows, it has persisted in the south-west. From 1975, badger setts were gassed in an attempt to stamp out the disease, but gassing was stopped in 1982; it was shown to be ineffective as well as inhumane. Badgers are now controlled by trapping.

In spring territorial disputes are frequent, and badly bitten boars may contract tuberculosis through infected saliva. In this way the disease can pass from group to group.

Young cows seem most likely to get tuberculosis. They are inquisitive and often approach a badger that is in their field, so may inhale the bacteria from an infected badger.

In farmland the badgers' territorial boundaries, marked by latrines, often follow hedgerows. Infected droppings on the border of cattle pastures may also be a source of cattle infection.

Infected badgers soon lose condition and start to behave oddly. They may become solitary and less wary of man. Attracted by cattle cake in troughs, some enter farm buildings. This may spread the disease.

Earthworms are plentiful in short pasture. Urine or saliva from an infected badger foraging for worms will contaminate the pasture.

The home and habits of a badger community

A badger community usually includes a number of adult boars and sows and one or two litters of cubs, up to 15 animals in all. Setts are often dug in sloping ground or under a rocky overhang, and are generally in woods or copses. Large setts may have more than 40 entrances and have been used for decades by generations of badgers.

There is usually one main sett and a number of outlying setts distributed around the territory. The main sett is a network of tunnels and chambers on several levels, usually within about 40 in. (100 cm) of the surface. Entrance holes are about 12 in. (30 cm) wide. Territories vary in size, but may be 100–125 acres (40–50 hectares) in extent.

Badgers mate between February and October, but the peak season is in spring. However, implantation of the fertilised egg is delayed until December and cubs are born from mid-January to mid-March, usually two or three in a litter. They are covered with a silky, greyish-white fur and are blind until about five weeks old. Weaning starts at 12 weeks old. Some cubs stay with the family group, others leave to find new territories and may move long distances to do so. Most of the cubs that leave do not go until their first winter, but some go in late summer when only a few months old.

In autumn, badgers lay down a large amount of fat under their skin, increasing their weight by up to 60 per cent. They do not hibernate, but from mid-December to mid-February activity is reduced and they live mainly off their fat.

Badgers often play together and groom each other. Mutual grooming may involve two badgers or several.

Elder bushes and nettles often grow near setts. Badgers eat elderberries and disperse seeds in their droppings. Soil enriched with dung favours elder and nettle growth.

Fallen trees near the sett provide both a playground and a food source. Badgers like to climb along them and to extract beetles, slugs and snails from under rotting bark.

Outside each sett entrance there is a mound of excavated soil, compacted from years of use. Old bedding may also be left outside the hole.

Half-grown cubs enjoy long play periods around the sett, chasing and jumping on each other. Adults often join in.

A scratching tree has sets of parallel claw marks roughly about $\frac{5}{16}$ in. (8 mm) apart.

The cubs are born in a lined breeding chamber, where they remain for about eight weeks.

Close to the main sett there is usually at least one latrine – a number of dung pits covering an area of several square yards. Often the ground around has been scratched up by badgers.

49

Most dives last less than a minute, but an otter can stay under water for as long as four minutes, and can swim at least 430 yds (400 m) without surfacing.

Food is usually carried ashore and is mostly small prey held in the teeth. The occasional large fish is killed and clasped to the chest.

Male

Female

The male, or dog, otter has a heavier head and thicker neck than the female (bitch) otter. She is also slightly smaller.

Although seen mostly in or near water, the otter may be sighted as it crosses open land, running with a hunched, rolling gait.

An otter eats with its prey held in its forepaws. Fish are its main food – eels are a favourite in many areas. It also eats frogs, mammals and waterside birds.

A swimming otter makes a wide, V-shaped wake, only its head showing. It swims smoothly, its forelegs tucked up, but it paddles with them to manoeuvre or gain speed.

Small
ears

Streamlined for speed in
the water, the otter has
small ears, a long body
with a powerful, tapering
tail and short, strong legs
with webbed feet. It often
stands upright to look
around, balancing on its
hind feet and tail. Male
36 in. (90 cm) head and
body, 16 in. (40 cm) tail.

Otters are playful. They often chase each other and
pretend to fight, especially when courting. While chasing they may make loud chirping noises.

Ears, eyes and nostrils set well
to the top of the otter's head aid
surface swimming. Its broad,
flat head helps to distinguish it
from a mink.

Scarce, except in Scotland.
Populations slowly
expanding in north and west.
Now in scattered places in
south and east.

Thick, tapered
tail

Webbed
feet

Otter *Lutra lutra*

Few people ever see a wild otter in Britain. Otters are rare today,
but until the 1950s they could be found throughout most of the
country. Since then numbers have declined rapidly. Even where
they do occur, it is more usual to see an otter's droppings than
the animal itself, except perhaps on Scotland's west coast.

Otters live by undisturbed waters where there is plenty of
cover, mostly by freshwater lakes, rivers, or quite small streams, as
well as some coasts. Fish are their main food. Strong swimmers,
as much at home in the water as on land, otters have large lungs
that aid underwater swimming. When diving they can slow
down oxygen consumption by reducing their heartbeats. They
can focus their eyes to see as well under water as on the surface
and have a moustache of stiff whiskers to help them feel their
way at the bottom of a muddy stream or in the dark; it may also
help in detecting prey.

No one knows for certain how long otters live in the wild,
but captive otters have lived to be 20. Humans have been the
otter's chief enemies, through hunting for sport and killing for
fur or for fish protection. Otters are now fully protected and are
slowly increasing in both numbers and distribution.

Once their fluffy coats have changed to waterproof coats, the bitch teaches the cubs to swim. They often have to be encouraged – even pushed in.

Families often play together, on land or in the water. Lone otters sometimes toss pebbles and catch them in their mouths.

Coastal otters and otter families

Only on parts of Scotland's western coast and islands are otters likely to be frequently sighted. Their numbers are greater there than in any other part of Britain, for the area is relatively undisturbed and food is fairly abundant.

Whether found by coastal or inland waters, dog and bitch otters live separately, coming together only for mating, which takes place at any time of the year. A bitch is on heat for about two weeks every 30–40 days. Courting otters usually find each other by scent – or sometimes by whistling to get each other's bearings. Before mating they often chase each other playfully and pretend to fight, both on land and in the water. The bitch is pregnant for about 62 days. The breeding den, or holt, is usually built in a quiet part of the bitch's territory. The two or three cubs – occasionally four or five – are born blind and toothless; weaning starts at about seven weeks old. They are two or three months old before they develop an adult coat.

The family breaks up when the cubs are about a year old, and the female begins to go on heat again. The cubs probably stay on the mother's territory for another few months before leaving, but may take several months to find permanent territories.

Oil spillage may affect coastal otters. Oil can mat an otter's fur and rob its coat of its natural waterproof protection, so it may die of cold.

Frequent grooming keeps an otter's coat sleek and waterproof. Its thick underfur traps an insulating layer of air against its skin, and is kept dry under water by the long outer fur.

An otter's holt is usually well hidden, often in a bank under overhanging tree roots. Holts in Scotland may be in more open places such as rocky cairns beside rivers, lochs or the sea.

A litter of generally two or three cubs is born in a holt lined with grass or moss. Cubs have fine grey fur at birth. Their eyes open when they are from four to five weeks old.

On the remote and rocky west coast of Scotland, otters are frequently seen by day. Here they feed on crabs, molluscs and various sea fishes.

53

Otter havens are quiet, specially protected stretches of water with plenty of bank vegetation. A dog otter's territory may cover a riverside strip about 12 miles (19 km) long, or 2–3 sq. miles (about 5–8 sq. km) of lake.

Havens to encourage the otter's return

Even where food is abundant, every otter needs a territory of several miles of undisturbed waterside with plenty of undergrowth, so otters have never been numerous in Britain. Since the 1950s riverside habitats have been extensively altered by building and farming developments as well as by clearing for drainage. Water sports have also increasingly disturbed waterways. All these have probably contributed to the otter's decline.

However, investigations during the 1960s suggested that the main reason was the increased use of pesticides, aldrin and dieldrin particularly. They cause water pollution and the contamination of fish with minute amounts of poison. The fish are probably unaffected, but the poison gradually accumulates in an otter eating a lot of them. A dog otter may eat 2–3 lb (1–1.3 kg) of fish a day. Following withdrawal of the most dangerous pesticides, otters have begun to increase again. However, in parts of central and southern England waters are still too polluted to provide a safe home. There is no evidence that the rapid increase of mink has led to the otter's decline.

Otter havens may help to bring back the otter to lowland Britain. Numbers of havens have now been established by landowners in co-operation with conservationists. Artificial holts are sometimes built where natural cover is sparse.

Scattered food remains such as fish scales mark the place where an otter has eaten a large fish.

54

A steep snowy or muddy river bank may be used as a slide by an otter family. The otters tuck up their forepaws and slide on their chests at speed into the water. They all repeat the game many times over.

Otter holts (dens) and lying-up places (which are sometimes called hovers) are usually under riverside tree roots.

Otters catch the prey most readily available, generally coarse fish 4–6 in. (10–15 cm) long. Large trout and salmon move too fast.

The spraints (droppings) are used to mark territory, and left in conspicuous places such as on rocks and fallen trees.

IDENTIFYING STOATS, WEASELS AND THEIR RELATIVES

The weasel, stoat, pine marten, polecat, mink and otter are all members of the weasel family (Mustelidae), and resemble each other in build, colouring or behaviour. All are very active hunters with long, sinuous bodies and comparatively short legs. They often sit upright on their haunches to look round, and when moving fast they generally bound along with the back arched. Males are usually up to 50 per cent bigger than females.

Confusion is most likely between the squirrel-sized stoat, especially a young one, and the smaller weasel. It is possible to confuse the pine marten, polecat and mink (all three roughly cat size), and perhaps also a swimming mink and swimming otter – the otter is much larger but its size is not obvious in the water. But each has certain distinguishing characteristics, and usually can also be identified from its habitat, although the stoat and weasel are found in most types of country. The mink and otter are likely to be seen in or near water, the polecat on farmland or lower hill slopes (mostly in Wales or the border countries) and the pine marten only in remote northern forests or moors. Except for the stoat and weasel all are fairly uncommon – the pine marten and otter are rare – and are active mainly at night.

Polecat identification can be confusing because of the existence of polecat-ferrets (descendants of feral ferrets), whose colouring can vary from pale to dark.

Creamy-yellow throat

Bushy tail

Pine marten
Martes martes
Head and body 18 in. (45 cm).
Page 59

Dark mask

Black legs

Polecat
Mustela putorius
Head and body 14 in. (36 cm).
Coat paler in winter.
Page 61

No distinct mask

Polecat-ferret
Page 61

Legs may be dark or pale

Pink eyes

No mask

Ferret
Page 61

Pale legs

Pastel mink
May be pale grey, buff or white.
Page 63

Stoat
in ermine.
Page 65

Stoat
Mustela erminea
Head and body 10 in. (25 cm).
Page 65

*Glossy
coat*

Black tip to tail

*Crisp line between
brown and white*

Mink
Mustela vison
Head and body 16 in. (40 cm).
Looks black at a distance.
Page 63

*Brown
throat-
patches*

*Short tail with
plain tip*

Weasel
Mustela nivalis
Head and body 8 in. (20 cm).
Page 67

*Irregular line
between brown
and white*

A swimming
mink has a
narrow head
and swims jerkily.

A swimming
otter has a
broad head
and swims
smoothly.

Broad flat head

Otter
Lutra lutra
Head and body 36 in. (90 cm).
Page 51

Heavy, tapered tail

Webbed feet

57

The pine marten is one of the few predators agile enough to catch a squirrel. If it falls, its lithe body twists to land safely on all fours from as high as 65 ft (20 m).

Food is obtained mostly on the ground and consists mainly of small birds and mammals which it seeks out by sight and smell. It also eats beetles, caterpillars, carrion, eggs and lots of berries.

The marten climbs trees with ease. It grasps the trunk firmly, digs in its claws, and uses both hind legs together to force itself upwards in a series of jerky movements.

Active mainly at night, the pine marten is rarely seen in daylight. It may be glimpsed as it crosses an open space such as a forest ride, and recognised by its distinctive bounding gait and bushy tail.

Although it can swim, the marten prefers not to. It leaps from stone to stone to cross a mountain stream.

Male

Creamy-yellow
throat

Female

The rare pine marten makes its home mainly in the remote forests and hills of northern Britain. It might be seen in a conifer forest at dusk or dawn.

Pine marten *Martes martes*

Once the lithe pine marten was widespread in Britain. Now it is uncommon, found mainly in remote forests or sometimes on rocky moorland where it spends quite a lot of time on the ground looking for food in forest rides and grassy areas. Trapping for its rich fur and persecution by gamekeepers led to its decline over the past 200 years, although it was never a serious threat to game birds. Felling of forests has also reduced its numbers, but today new plantations offer it a chance of expansion.

Pine martens breed only once a year, mating in July or August. But there is a delay in the implantation of the fertilised egg, and females do not become pregnant until about January. A litter averages three babies, born in March or April in a den usually in a crevice among rocks or tree roots. The young spend at least six weeks in the den before their eyes open and they are big enough to venture out, and the family stays together until they are six months old. Youngsters grow quickly, reaching adult size in their first summer, but until their first winter moult they have paler, woollier fur than adults. Apart from man, the pine marten has no serious enemies and can expect to live for several years – some may reach ten or more.

Found in remote areas, but it is spreading into new forest plantations.

The pine marten, alert and elusive, is cat-size, with noticeably long rich brown fur and a bushy tail. Its conspicuous creamy-yellow throat is distinctive. Male 18 in. (45 cm) head and body, 9 in. (23 cm) tail. Female slightly smaller.

59

The polecat preys mainly on small creatures such as voles, mice, frogs and worms, but will also chase and catch animals as large as hares. It may sometimes be seen hunting during the day.

Litters contain from five to ten young, born with white silky hair. Darker fur grows later, and the young leave the nest when two months old, about August. Family groups may be seen together in late summer, but polecats are normally solitary.

Summer

Winter

In summer the woolly underfur is largely hidden by the guard hairs. In winter the denser underfur makes the guard hairs stick out, so the animal looks paler and rounder.

The pea-sized stink glands on the underside, at the base of the tail, secrete a persistent, foul-smelling scent, used defensively or to mark a territorial boundary.

The male polecat is slightly larger than the female. Mating occurs between March and May; courtship is rough, the male dragging the female by the scruff of the neck. Usually there is one litter a year.

Dark
guard hairs

Dark
mask

Polecat families can be quite sociable, and they will indulge in mutual grooming. The paler animal of the two (left) is probably a polecat-ferret.

The polecat has creamy-yellow woolly underfur protected by long, dark guard hairs, and may vary in colour from pale to almost black. A polecat-ferret looks similar, but its face markings differ. Male 15 in. (38 cm) head and body, 5½ in. (14 cm) tail.

Found mainly in Wales but spreading. Polecat-ferrets found throughout Britain.

Polecat   Polecat-ferret

Ferret

A polecat has white ear tips and a dark mask. A polecat-ferret usually has a paler forehead and no mask. A pure ferret is creamy-white.

Polecat *Mustela putorius*

Only in central and western Wales did the polecat survive the centuries of persecution that led to its near extinction in Britain. Once known as the foul-mart because of its strong smell, it used to be widespread, but was ruthlessly trapped and killed for its fur (known as fitch) and because it was considered a threat to game and livestock. Since the 1950s trapping has declined and numbers are increasing, but the picture is confused by the numbers of ferrets living wild in many parts.

Ferrets are creamy-white, domesticated polecats used to catch rabbits, and some escape or are lost. Over generations, many have reverted to polecat colouring and are known as polecat-ferrets; there are established colonies in many counties, and they have also interbred with the spreading population of true polecats. So polecats tend to vary in colour because their long outer (guard) hairs may range from creamy-yellow to almost black, depending on how much ferret and how much true polecat is in their ancestry. Polecats are commonest on farmland and are mostly active at night. They often inhabit abandoned rabbit burrows. Some live to be five or more. They have few natural predators but are often killed by cars.

61

A swimming mink can be distinguished from a swimming otter by its more pointed snout, darker colouring and smaller size. It has a longer body and thicker tail than a water vole.

One litter of five or six young mink is produced a year. They leave the den at about two months old.

All wild mink in Britain are descended from animals that escaped from fur-farms. Descendants of those specially bred to provide pale-coloured pelts may still be seen, and are known as pastel mink.

The mink catches more of its food on land than the otter does, and is more likely to be seen out of the water in pursuit of rabbits, voles and other small creatures.

Mink swim well. Some of their food is caught in the water. Usually they take slower-moving prey such as eels and crayfish.

Chocolate-brown fur

When a mink leaves the water its wet fur looks black and spiky. It shakes off excess moisture and may sit on the bank to groom and clean its coat.

Mink *Mustela vison*

When the mink was introduced to Britain from North America in the late 1920s, it was intended to be kept captive on fur farms and raised for its valuable pelt. But many farms had inadequate fences and, as mink are good climbers, many escaped and bred in the wild. Since 1930 they have spread over most of Britain.

At first mink were assumed to be a pest and attempts were made to get rid of the wild population. Thousands were trapped but with little noticeable effect. The mink now appears to be a permanent addition to Britain's wildlife, with few threats to its continued existence. At a time when many British mammals are becoming rare, a successful newcomer might seem a welcome addition, but there is still controversy over the mink's acceptability. Mink eat a lot of fish, and their depredations among trout on a fish farm or valuable young salmon can be serious. They also eat birds and water voles (a major factor in the decline of this species). Although mink probably do not seriously compete with otters for food, they may prevent otters from recolonising suitable habitats. Young mink are born in a den among waterside stones or tree roots. From June onwards they can be seen foraging with their mother, and are fully grown by autumn.

Rarely seen far from a river or lake, the mink is mostly active at night, often preying on waterfowl. Its dense, glossy, chocolate-brown fur looks almost black from a distance, when wet especially. Male 16 in. (40 cm) head and body, 5 in. (12.5 cm) tail. Female usually smaller.

A waterside animal becoming steadily more numerous and still extending its range.

The sexes look alike, but a full-grown male is up to 50 per cent bigger than a female. Pairs may be seen together briefly in summer, at breeding time.

Male

Female

In northern Scotland stoats commonly turn creamy-white in winter, and are then called ermine. The colour change may take only a few days. Elsewhere, except in very cold weather, stoats may go partially white and have a patchy look. In southern England and in Ireland, stoats often stay brown.

Even if the stoat turns fully white in winter, the tip of its tail remains black.

The stoat can move fast, perhaps up to 20 miles (32 km) an hour in bounds of 20 in. (50 cm). It may be seen in all types of country, but usually where there is good cover. In farmland it keeps mainly to hedges, walls or fences.

There is one litter of six or more young a year. The young first leave the den at about five weeks old, and often hunt and play in a family group.

*Creamy-white
underparts*

The stoat often kills prey more
than twice its own size by biting
deeply into the neck. Rabbits,
small mammals and birds
provide the bulk of its diet.

*Black
tail-tip*

Sand-dunes are often frequented by rabbits, so are a
good hunting ground for stoats. Once they start in
pursuit, the prey has little chance of escape.

Alert and inquisitive, the stoat
often sits upright to view its
surroundings, revealing the
sharp division between its
creamy-white underside and
brown flanks. Its black tail-tip is
distinctive. Male about 10 in.
(25 cm) head and body, 3 in.
(76 mm) or more tail.

Widespread in woods, farms,
uplands, despite extensive
trapping and shooting.

Stoat *Mustela erminea*

Slim and savage, the stoat is one of the fiercest of predators,
active by day or night. It relentlessly tracks down its prey by
scent, and its habit of sometimes licking blood from its victim's
fur has given rise to the legend that it sucks blood.

Although the stoat has suffered extensively at the hands of
gamekeepers, it has remained widely distributed and numerous.
The biggest threat to its existence was the introduction of
myxomatosis in the 1950s, the disease that wiped out almost all
the rabbit population. But as the stoat eats many other things
beside rabbits, even occasional insects, it was able to survive in
most parts. A stoat's hunting ground usually covers about 50
acres (20 hectares). Its den is in a rock crevice or abandoned
rabbit burrow, and it normally lives alone. Stoats mate in
summer, but implantation of the fertilised egg is delayed until
the following March, females giving birth in April or May.
Young stoats, independent at about ten weeks old, may be
mistaken for weasels but can be recognised by the black tail-tip.

In Ireland, stoats are darker and, like weasels, have a wavy
dividing line between flank and belly fur. As there are no
weasels in Ireland, there is no problem with recognition.

Although a weasel hunts mainly by scent, investigating all likely holes and crevices, it will attempt to seize a small bird such as a meadow pipit if it disturbs one into flight.

Young weasels are born in a nest of leaves or grass in a hole or crevice. A litter usually numbers five or six. Their eyes open when they are about three weeks old.

Voles and mice are the main food, along with a few rats and rabbits as well as small birds and their chicks if the chance arises. Prey is killed by a bite at the back of the neck.

Prey is usually taken on the ground, but weasels climb well and sometimes raid a bird's nest-box. Their very slim bodies can squeeze through an opening only $1\frac{1}{8}$ in. (28 mm) across.

Brown tail

Wavy flank line
The weasel's brown fur meets its white underparts in an irregular line along its flanks, and there are small brown patches on its throat. Its tail is brown to the tip. Unlike the stoat, its coat does not go white in winter. Male about 8 in. (20 cm) head and body, 2 in. (50 mm) tail. Female slightly smaller.

Brown throat-patch

Widespread in most habitats, but absent from Ireland and many of the smaller islands.

A weasel is most likely to be seen as it streaks across a road, its slim body fully stretched and its short legs moving so fast they blur.

A weasel often sits upright on its haunches to look around, showing up its wavy flank line. No two weasels have exactly the same flank pattern.

Weasel *Mustela nivalis*

The weasel looks something like a long, slim, fast-moving mouse, and often moves in undulating bounds of 12 in. (30 cm). It is the smallest British carnivore, and a fierce hunter by day or night. Mice are one of its main foods – one weasel may eat hundreds in a year, so farmers and foresters should regard it as a friend. Each weasel has a territory of 10–20 acres (4–8 hectares), females using much smaller ones than males. Territory size depends on the food available; where there is plenty there is no need to hunt so far afield. A weasel eats roughly 1 oz (28 g) of food a day – about 25 per cent of its own weight.

Young weasels are born in April or May, and there may be a second litter in July or August. Youngsters stay with the mother, often hunting in family parties, until up to 12 weeks old. By this time they are fully grown. Unlike other British carnivores, which do not breed in their first year, young weasels may be capable of breeding during their first summer.

Most weasels do not live to be more than a year old. Gamekeepers trap large numbers as vermin, and many are killed on the roads. Cats, owls, foxes and birds of prey will also kill weasels, but risk a hard fight in doing so.

67

The pup can swim almost from birth and may go to sea on the next tide. Pups suckle on land or in water for a month or more.

Common seals mate at sea in early autumn. A female (cow) has only one pup a year, usually born in June or July on an exposed rock or sand-bank.

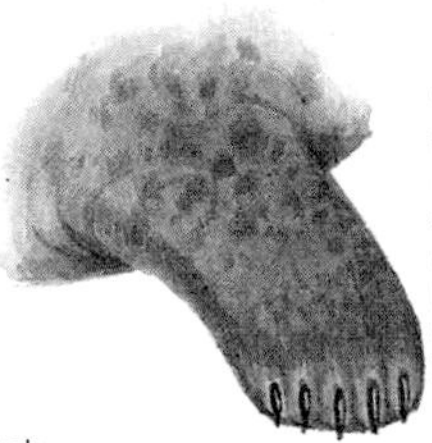

The fore limb is a flipper with five long black claws, normally held close to the body. It acts as a stabiliser in the water.

When wet, the seal looks much slimmer and almost black. On land it moves with an awkward, humping motion, hardly using its fore flippers and not using its hind ones at all.

Fish are the main food, caught mostly by diving to search the sea-bed. Flatfish and other bottom-dwelling fish are taken. Young seals may eat shrimps.

Rounded head

Short muzzle

A common seal often basks with body arched so that its hind flippers and small, rounded head are raised. Seals vary in colour but are mottled with dark spots. From a distance they look pale when dry. Male 68 in. (173 cm) long, female smaller.

Mottled coat

As land travel is hard work for common seals, they tend to gather in shallow water over sand-banks and let themselves be stranded by the outgoing tide.

Common seal *Phoca vitulina*

Although common seals are often seen in groups of more than 100 basking contentedly in the sun, they are not really social animals. Unlike grey seals, they do not live in organised herds, nor are they noisy, even at breeding time. Common seals are placid, and tend to be found on sand-banks around sheltered shores such as estuaries and sea lochs.

Common seals will, however, haul themselves on to seaweed-covered rocks close to deep water to bask, often returning to the same place day after day and departing to feed only when the incoming tide displaces them. On sand-banks they allow the outgoing tide to strand them on the highest point, a good look-out position where they may stay for several hours. If disturbed they move close to the water's edge, ready for a quick getaway. They have good eyesight, on land or under water and unless they are asleep it is not easy to approach them on open sand-banks. About one-third of the British population live around The Wash but they travel considerable distances along the east coast, probably following fish shoals. They may live for 20-30 years. An outbreak of viral disease that began in 1988 severely depleted their numbers, but the population has now recovered.

May be seen off south coast but does not breed there. Rare off west of England, Wales.

In the water the seal's very rounded head looks like a buoy or fishing float. Its short muzzle and V-shaped nostrils help to distinguish it from a grey seal, which has nostrils that are almost parallel.

69

The top of the seal's head is flat. Grey seals are often very noisy, barking, hooting, moaning, hissing and snarling, on land and in water.

Cow seals are paler than bulls, especially when their fur is dry, and not nearly as heavy – only about half a bull's weight. They are also shorter in the muzzle and thinner in the neck.

In fine weather, seals haul themselves onto rocks to bask, sometimes in large numbers. Flat-headed grey seals may be seen sharing a rock with smaller, round-headed common seals.

Where breeding seals overflow the beach onto soil, mud patches may be formed and seals wallow in them. Wallows soon become contaminated and a risk to the seals' health.

Grey seals eat larger fish than common seals. Their diet includes cod, whiting and salmon, and they will raid fishing nets and fish farms to take them.

Long, pointed head

Rolls of flesh

Year after year seals use the same isolated rocks or shingle coves for basking. The sites are often different from the regular breeding places.

Grey Seal *Halichoerus grypus*

Grey (or Atlantic) seals may be seen off most of Britain's coasts but are most numerous around Scotland, particularly the offshore islands. They gather to breed on shore, sometimes in their hundreds, from September to December. Outside the breeding season the seals spend much of their time at sea, sometimes for weeks on end, but haul themselves onto rocks or the shore to bask from time to time. While moulting, about March, the seals rest for long periods on rocks above high-tide level. Newly moulted seals are easily spotted because of their bright coats.

Fish are the seals' main food, adults eating a daily average of about 22 lb (10 kg), although they do not feed every day. For most of this century seal numbers have steadily increased. Colonies on places like the Farne Islands have become very overcrowded, causing damage to the soil and a lot of disease among pups. Because of their large numbers and their fish consumption, and also because they carry parasites that are transmitted to fish, there have been attempts at regular licensed culling (selective killing). These aroused so much public hostility that grey seals are now mostly left alone, except in cases where individuals cause damage to cages of salmon in sea lochs.

Commonest around rocky shores, mainly off north and west coasts of Britain.

When dry, a male seal's fur is a dark, blotched grey or brownish-grey. When wet it looks almost black. The head is long and pointed and the body very portly, with thick rolls of flesh folding around the neck. Male (bull) about 7 ft (2·1 m) long, female (cow) about 6 ft (1·8 m).

Scavenging skuas patrol the breeding beaches looking for dead and sickly pups. These birds also feed on the afterbirth.

A cow returning from feeding finds her own pup by first recognising the place, then picking out its voice and finally by sniffing the pup to check its scent.

If the breeding beaches get overcrowded, some cows are driven inland. Their heavy bodies erode the soil and spoil the area for breeding puffins the following May – their burrows often collapse.

A pup sheds its long, creamy-white baby coat after two or three weeks and becomes pale grey. At two months old it is ready to go to sea.

A grey seal breeding colony

In the wet, misty days of autumn grey seals come ashore to their breeding grounds on rocky islands or lonely beaches – normally to the same place year after year. The big bulls arrive first to establish territories above the high-tide line. The biggest bulls, the beachmasters, have the most extensive territories and the largest harems – about ten or more cows – so they father the greatest number of next season's pups.

The cows come ashore a few days after the bulls and are ready to give birth from the previous season's mating. They join a harem, not necessarily with the same bull every year, and are jealously guarded while they have their single pup because they are ready for mating soon after giving birth. Implantation of the fertilised egg is delayed for about three months so that a cow will not give birth before the following autumn.

Once the pups are weaned and leave the breeding beaches, they do not return for several years until they are adult. Cows mature at four or five years old, then come ashore to breed every year until they are about 35 or more. Bulls may mature at about six years old but are not big enough or strong enough to gain and defend a territory until they are at least nine. They then breed every year for about four or five seasons, but become worn out with the strain. Few survive beyond about 20.

Late-arriving bulls may challenge beachmasters for possession in noisy contests of strength. Often they fail to secure a patch and are forced to stay on the fringe of the colony or in the shallows.

After her pup is three weeks old, a cow goes to feed as the tide rises. A bull dare not desert his territory, so lives off his fat reserves for nearly eight weeks.

A cow defends the area around her own pup and fiercely repels intruders, including other pups, with bared teeth, hoots and threatening lunges. Some pups become lost or are abandoned.

A cow bears one pup, which feeds on her rich milk every five hours for three weeks. During this time the pup triples its 32 lb (14 kg) birth weight, but the cow cannot go to sea to feed.

The tail and mane are thick. Around the tail root the hairs form a 'thatch' that sheds the rain.

Noted for its stamina and hardiness, the pony has strong, wiry hairs that shed the rain. The whorl-patterning of the hair on its flank throws the rain off on each side, away from the crease of the groin.

New Forest pony

Wild ponies have been known to roam the New Forest in Hampshire since Saxon times, and half-wild herds still remain. The ponies may be any colour except piebald and skewbald. There are two types: lightly built ponies up to 13·2 hands (54 in., 137 cm) high at the withers, and heavier-built ponies up to 14·2 hands (58 in., 147 cm).

Foals are born between March and November, most births occurring about April. Foals have short manes and tails that reach adult length after about two years.

Oatmeal-
coloured
muzzle

Small breeding herds roam free on Exmoor, each one led by a stallion. The winter coat is shed about May and grown again by September.

Strong and stocky in build, the Exmoor pony is short-legged and broad-chested with a broad, level back. The common colours are dun, bay and brown, all with distinctive oatmeal colouring round the eye and on the muzzle and underparts. Stallion up to 12·3 hands (51 in., 130 cm) high at withers; mare slightly smaller.

Half-wild herds on Exmoor, but many domesticated ponies found throughout Britain.

The face is broad with a wide forehead, short, thick ears, a deep jaw and broad, black nostrils. A distinct fleshy ridge over the eye (known as toad eye) deflects the rain.

Exmoor pony *Equus* (domestic)

There have probably been ponies living on Exmoor since about 60,000 BC, well before the last Ice Age ended. Today's Exmoor ponies, the toughest and truest to type of all Britain's native ponies, belong to the Northern Pony family, a type that can survive in Arctic regions. They are hardy and well equipped to survive cold and wet, living outdoors on Exmoor through the hardest winter. The thick, wiry coat covers a dense undercoat that insulates the body so well that in winter ponies often have unmelted snow lying on their backs, and their roomy nostrils have a large surface of skin that warms up the air before it enters the lungs. The ponies have deep jaws with strong, deeply rooted teeth for dealing with the tough moorland grass.

Exmoor ponies are not only hardy, they have great strength and endurance. Despite their small size they can easily carry a 12 stone (75 kg) man all day, and are used as mounts by Devon shepherds. Although breeding herds run free on Exmoor, the ponies are only half wild. Each one has an owner, and every October herds are rounded up and branded. Young animals may be sold and trained for driving in harness or as riding ponies. Their strength makes them useful for pony-trekking.

75

DEER IN SUMMER

Once they have moulted their winter coats – usually by May or June, depending on the species and the weather – deer have bright, glossy summer coats. Several species are similar in colour and build. The main identification points are the variations in rump pattern, tail markings and length and, in males, the pattern of the antlers. Within a species, both antler patterns and shoulder heights, which are given as a guide to species size, can vary with age and among individuals of the same age.

SMALL RED-BROWN DEER WITH SHORT ANTLERS

Roe buck
Capreolus capreolus

Antlers clean, no velvet. Black muzzle, white chin. Creamy-buff rump, no visible tail. Shoulder height 26 in. (66 cm). Page 95

Muntjac buck
Muntiacus reevesi

Antlers with velvet. Dark face stripes. Plain muzzle. Rump plain but shows white when tail lifted in alarm. Shoulder height 19 in. (48 cm). Page 99

SMALL RED-BROWN DEER WITHOUT ANTLERS

Muntjac doe
Muntiacus reevesi

Dark patch on forehead. Plain muzzle. Back often rounded. Tail inconspicuous but shows white below when raised in alarm. Shoulder height 18 in. (45 cm). Page 99

Chinese water doe or buck
Hydropotes inermis

Large, round ears. Black nose. Plain rump. Buck has two protruding, tusk-like teeth. Shoulder height 22 in. (55 cm). Page 101

Roe doe
Capreolus capreolus

Black muzzle, white chin. Creamy-buff rump; no visible tail, but has a hanging tuft of hair. Shoulder height 25 in. (64 cm). Page 95

SPOTTED DEER WITH ANTLERS

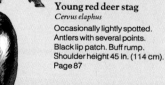

SPOTTED DEER WITHOUT ANTLERS

Fallow doe of typical colour
Dama dama

No lip patch. Long, black, white-fringed tail. White rump with black border. Shoulder height 30 in. (76 cm). Page 83

Sika hind
Cervus nippon

Black lip patch. White tail, often with a dark stripe. White rump with dark edging. Shoulder height 28 in. (70 cm). Page 91

Young red deer stag
Cervus elaphus

Occasionally lightly spotted. Antlers with several points. Black lip patch. Buff rump. Shoulder height 45 in. (114 cm). Page 87

Fallow buck of typical colour
Dama dama

Antlers with several points and broad blades. Plain lip. Long, dark, white-fringed tail. White rump with black border. Shoulder height 37 in. (95 cm). Page 83

Sika stag
Cervus nippon

Antlers with several points. Black lip patch. White tail, often with dark stripe. White rump with dark edging. Shoulder height 32 in. (80 cm). Page 91

Red deer calf a few months old
Cervus elaphus

Black lip patch. Fading spots. Buff rump. Shoulder height under 30 in. (76 cm). Page 87

DEER IN WINTER

Winter coats usually start growing about September and are duller and thicker than summer coats. Species that are spotted in summer generally lose their spots in winter. Most young deer change to adult colouring when they are a few months old. As in summer, the main identification points are the variations in rump pattern, tail markings and length and, in males, antler patterns. Shoulder heights are given as a guide to species size, but vary with age and among individuals.

GREYISH-BROWN OR BROWN DEER WITHOUT ANTLERS

Muntjac doe
Muntiacus reevesi
Back often rounded. Plain rump. Tail inconspicuous but shows white below when raised. Shoulder height 18 in. (45 cm). Page 99

Roe doe
Capreolus capreolus
Black muzzle. White rump patch with a short tuft of hairs hanging down. Shoulder height 25 in. (64 cm). Page 95

Red stag without antlers (hummel)
Cervus elaphus
A mature male able to breed although it has failed to grow antlers. Heavy mane on neck. Black lip patch. Buff rump. Shoulder height 46 in. (117 cm). Page 87

Fallow doe of typical colour
Dama dama
Long, black, white-fringed tail. White rump with black border. Shoulder height 30 in. (76 cm). Page 83

Black fallow doe
Dama dama
Long black tail; tip may be copper-coloured. No white on rump. Shoulder height 30 in. (76 cm). Page 83

Sika hind
Cervus nippon
White tail, often with a dark stripe. White rump with dark edging. Shoulder height 28 in. (70 cm). Page 91

Red hind
Cervus elaphus
Black lip patch. Buff rump. Shoulder height 45 in. (114 cm). Page 87

GREYISH-BROWN OR BROWN DEER WITH LARGE ANTLERS

Young red stag
Cervus elaphus
Antlers with several points. Buff rump. Shoulder height 43 in. (110 cm). Page 87

Sika stag
Cervus nippon
Antlers with several points. White tail, often with dark stripe. White rump with dark edging. Shoulder height 32 in. (80 cm). Page 91

Young black fallow buck
Dama dama
Antlers with several points but blades not yet broad. Long dark tail; tip may be copper-coloured. Shoulder height 35 in. (89 cm). Page 83

GREYISH-BROWN OR BROWN DEER WITH SHORT ANTLERS

Red stag
Cervus elaphus
Antlers with many points and two low forward branches. Buff rump. Shoulder height 48 in. (120 cm). Page 87

Muntjac buck
Muntiacus reevesi
Antlers clean, no velvet (except on immature bucks). Rump inconspicuous, shows white when tail lifted in alarm. Dark stripes on face. Muzzle plain. Shoulder height 19 in. (48 cm). Page 99

Fallow buck of typical colour
Dama dama
Antlers with several points and broad blades. Long, dark, white-fringed tail. White rump with black border. Shoulder height 37 in. (95 cm). Page 83

Young sika stag with first antlers
Cervus nippon
Antlers clean, no velvet. White rump with hairs fluffed as alarm signal. Plain muzzle. Shoulder height 28 in. (70 cm). Page 91

Roe buck
Capreolus capreolus
Antlers with velvet. White rump with hairs fluffed as alarm signal. Black muzzle. Shoulder height 26 in. (66 cm). Page 95

Deer and their antlers

Only deer grow antlers, which are quite different from horns although they are used similarly – mainly as weapons. Horns are permanent bony growths covered with a sleeve of horn, and occur in both sexes. Antlers are bony growths that drop off and are regrown every year; except for reindeer, only males grow them. While antlers are growing they are covered by a hairy skin called velvet. Blood vessels in the velvet supply food and oxygen to the growing bone. When the antler is fully grown the velvet is shed or rubbed off and the antler dies, although it remains on the deer's head for several months.

The pattern of the antler cycle is similar in all species, but times of casting and cleaning differ. Most species in Britain shed their antlers at some time in early spring or summer and have new ones fully grown in late summer or autumn, before rutting time. Roe deer, however, shed their antlers in winter and have fully grown ones about April. With some species, each pair of antlers grown is normally larger and with more tines, or points, than the previous pair until the deer reaches old age. A deer's age cannot be measured by the number of points it has, but they do indicate whether it is young or old.

The antler cycle of a fallow deer

FIRST PAIR

March

May

1 Growth begins
Early in the year after its birth, a male fallow fawn grows pedicles, or stalks, for its antlers.

2 Antlers appear
Soon the pedicles bear small antlers covered with a hairy skin called velvet.

SECOND PAIR

Late May

Late May

Mid-June

Early July

Late July

6 Antlers cast
The antlers are cast, or shed, one at a time. The second antler may be cast a few days after the first.

7 Scabs form
When an antler is cast, blood oozes slightly from the pedicle, then dries up and a scab begins to form.

8 New growth begins
Two weeks later the antlers are growing fast, and show clearly above the pedicles.

9 Branches form
Soon each antler branches into a brow tine, pointing forwards, and a main beam, pointing backwards.

10 More tines appear
The antlers grow rapidly, and more tines (or points) branch off the main beam.

Mid-May

Late July

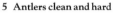

Mid-August

3 First pair growing

Growth is rapid, but size varies. The first set may be short stubs or slender spikes. Rarely they may have one or two branches.

4 Growth complete

When the antlers are fully grown, the velvet dries and shrivels. The buck rubs it off against a tree.

5 Antlers clean and hard

When all the velvet has been cleaned off, the antlers die but stay on the pedicles until late the following spring.

Late August

Late August

End of August

11 Growth complete

A month later the antlers are fully grown but are still covered in velvet.

12 Velvet shed

As the velvet is shed and rubbed off, the antlers are untidy and blood-stained for several days.

13 Antlers clean

A few days later all the velvet is gone and the antlers are clean and hard. In the following spring the cycle begins again.

At rutting time, a buck's neck is enlarged and his Adam's apple even more conspicuous. He utters loud, bellowing groans that sound eerie and travel far. Hollows scraped in the ground are scented with urine to mark his territory.

Fallow does have no antlers. Menil deer retain some spots on the grey-beige winter coat that begins to appear in late September or October. Others are seldom spotted in winter.

Fawns are born singly in June, usually hidden in long grass or bracken. Black fawns have brownish dappling. Typical colouring is chestnut-brown with white spots. Unspotted sandy fawns grow up to be white animals.

Menil fawns are pale brown with white spots. All menil fallow deer have brown rather than black markings on the back, rump and tail.

At rutting time a buck rubs his head against saplings to mark his territory, and frays the bark. He also thrashes his antlers against branches and bushes.

Most fallow deer are in full summer colouring by mid-June after a moult of about 40 days. The long, black, white-fringed tail and white, black-bordered rump are distinctive. The buck has broad-bladed (palmate) antlers and a prominent Adam's apple. Buck about 37 in. (95 cm) high at shoulder; doe slightly smaller.

Broad-bladed antlers

Long tail

Typical summer coat

Fallow deer like to live where there is grass for grazing and mixed or deciduous forest with shrubby undergrowth for shelter and browsing.

Fallow deer *Dama dama*

Wild herds of fallow deer have lived for centuries in ancient forests such as the New Forest, Epping Forest and the Forest of Dean. They were a favourite quarry of medieval huntsmen, and later became the first choice for gracing the parks of hundreds of stately homes. During this century, animals that have escaped from parks have established many feral herds.

The fallow is unusual among deer because there are several colour varieties. Chestnut–brown with white spots is the typical summer colouring, but some animals, known as menil, are pale brown with spots, and there are intermediates between the two as well as both black and white varieties. Park herds may be all one colour but in the wild mixed herds are more usual.

Fallow deer may be seen feeding at any time, but dawn and dusk are generally favoured in the wild. By day they rest and ruminate in undergrowth or undisturbed pasture. Rutting time – the October to November mating period – is the most exciting time to watch them as the male (buck) herds together a group of females (does). Rival bucks fight fiercely, often with much clashing of antlers and furious charging until one retires hurt or defeated and the other takes the harem.

Most wild herds in south and east. Common in parks all over Britain.

Even in bucks of the same age, antlers vary in size and shape. The three bucks shown are all in their fourth year. A mature buck has antlers up to about 20 in. (50 cm) long.

83

Summer in a fallow deer park

Fallow deer are a pleasure to watch at any season, and one of the best places to see them is in a deer park, particularly in late June when most of the fawns are old enough to gambol together in the cool of the evening. Fawns spend much of the day trotting after their mothers and grazing, but are suckled several times a day. Many does suckle their fawns into the new year.

The bucks cast their antlers at any time from late March to early June. They spend summer in a bachelor group while new antlers are developing. When the antlers are fully grown – towards the end of August for older bucks – the soft covering, or velvet, is cleaned off by rubbing them against trees until they are clean and hard, ready for rutting in autumn. In winter, many park deer are provided with extra food, usually hay, which is placed in racks, or root crops such as swedes, which are spread on the ground.

From two years old a doe normally bears a fawn every year for perhaps ten years or more, and herds increase rapidly. To prevent overcrowding and disease and the destruction of their habitat, the deer are regularly culled (selectively killed); bucks from August to April, does from November to February.

Grazing does keep a watchful eye on their fawns playing near by. The deer constantly flick their tails to brush away flies.

Animals of different colour varieties interbreed. A fawn is not necessarily the same colour as either parent.

Most fawns are born about mid-June, and by the end of June are two or three weeks old. They sometimes play in groups, chasing each other and jumping on and off grassy hillocks.

The bucks live peaceably together in summer, separate from the does and fawns. They put on weight as they graze on the rich grass, and are in their prime in August and September – the 'fat buck' season.

Trees have a distinct browse line at about 5–7 ft (1·5–2·1 m) – the height the deer can reach to feed when standing on hind legs. Palatable shrubs such as hawthorn are often strangely shaped by deer browsing.

Young trees are protected by wooden or metal cradles to prevent deer browsing on shoots and stripping bark.

Late-born fawns may be seen resting in the grass while others play. All fawns rest for much of the day in the first week or two of life.

Deer seem to enjoy the taste of a salt lick, fixed to a tree or stump. It provides them with minerals such as sodium, calcium and magnesium.

The deer use either their feet or antlers to scratch when irritated by parasites. Stags grow new antlers every summer. Antlers are full-grown and hard by September.

Hinds are slightly smaller than stags, and lack antlers. A deer grooms its coat by licking it and combing it with the front teeth of the lower jaw.

Family groups comprising a hind, her calf and a yearling born the previous summer are frequently seen together. The deer are often active at dawn and dusk.

A young stag's first antlers are usually simple spikes. They are seldom cast before June and sometimes as late as September, long after those of mature stags, which are cast in March or April.

Calves are brown with white spots and are well camouflaged against a background of heather, bracken or moorland grass.

Calves are born singly, most in late May or June. A hind cleans under her calf's tail as she suckles it. The calf lies hidden in the undergrowth for much of the first week or so of its life.

Many-tined
antlers

Buff
rump

Originally woodland animals, red deer in Scotland have adapted to live on moors and open hillsides because so much forest has been felled.

Red deer *Cervus elaphus*

Scotland is the stronghold of Britain's wild red deer – the largest of our native animals. There are probably more than a quarter of a million red deer in the Highlands and Islands. Most other wild herds are on Exmoor and in the Quantock Hills, the New Forest, Thetford Forest and the Lake District.

Except at rutting time in autumn, red deer stags (adult males) live in separate groups from the hinds (females). Deer in woodland live in small groups, sometimes resting and ruminating by day and emerging to graze at night. Highland deer are usually seen in larger herds, moving up the hillsides by day to feed on grasses, heather and lichens and sheltering in the deeper heather or woods at night. In summer they usually keep to the higher hill slopes, following the growth of new heather. In winter they move to lower ground. The red deer's bright summer coat changes to a thicker grey-brown from about September to May. Although Highland deer are hardy and can scrape through the snow to find food, many die in severe weather. The farming of red deer for venison has become widespread in Britain. The deer are kept in a large enclosure and given hay and other foods to help them through winter.

Biggest wild herds found in Scotland, Devon, New Forest, Cumbria. Common in parks.

Britain's native red deer takes its name from the bright red-brown of its summer coat. It has a short tail, usually held close against its buff-coloured rump. Mature stags have antlers with many points (tines). Antlers can grow to about 28 in. (70 cm) long. Stag about 48 in. (120 cm) high at shoulder.

For most of the year the stags live in separate groups from the hinds, yearlings and calves. In September the stags split up, each stag seeking out the hinds to try gathering a harem.

Rutting stags emit deep, loud, bellowing roars. Rivals try to outdo each other's roaring. They assess each other from the performance before they commit themselves to a fight.

A stag will wallow in the mud until its body is mud-plastered. This helps to spread its strong rutting smell all over its body.

A stag is usually about five years old before it can successfully keep a harem. Young males that linger near a harem are chased off by the stag in possession.

The roaring of the red stags

For most of the year red deer live quiet and unobtrusive lives, but for about three weeks beginning in September, the countryside echoes to the stags' bellowing roars and the clashing of antlers. This is the rutting season, the courting and mating time when the stags move into the areas where the hinds live. Stags are in their prime at the start of the rut, with hard, fully grown antlers, a thick neck and a heavy winter mane as well as plenty of fat reserves built up during the rich summer's grazing.

Each stag tries to round up a harem of hinds, but his success depends on his size, age and how impressive he looks to other stags. The more he is threatened by rivals, the more he displays his strength by bellowing or thrashing trees with his antlers. Generally stags try to avoid fighting. Although intense and prolonged fights do occur between evenly matched animals, most contests are settled by a display of strength. The most successful stags are those about eight years old; they may hold a harem of 10–20 hinds. Stags under five are rarely strong enough to hold together a harem, and those over 11 are too old. Rutting stags use up an enormous amount of energy and hardly have time to eat, so often emerge from the rut thin and exhausted.

As each hind in the harem becomes ready for mating, she emits a scent that attracts the stag, which may chase and sniff her.

Rival stags often walk beside each other, slowly and some way apart, while they assess each other's strength. One may walk away, but if they appear to be evenly matched they will begin to fight.

Before mating, the stag licks and nuzzles the hind. A southern hind is usually two years old when she has her first calf; a hind in the harsher Highlands may be three or four.

Stags fight with a furious clashing of antlers as each tries to push away the other. Injuries or broken antlers may result, but fights to the death are rare. Very occasionally two stags cannot unlock their antlers and both die.

During the rutting season the deer are moulting from summer to winter coats. A stag grows a mane that is present all winter.

If one or both rival stags have cast their antlers (in April or May), they settle a dispute by boxing instead of clashing antlers.

When alarmed, a sika deer flares its rump hairs so that the patch is very conspicuous. It acts as a warning signal to others as the deer flees from danger.

Calves are born in May and June. They have white spots at birth, but after a few months the spots disappear.

Sika deer can live in coniferous woods but prefer mixed woods with areas of shrubby undergrowth. They will graze on nearby grassland.

When first alerted to danger, a sika deer will turn to face the disturbance before deciding whether or not to run away.

The hind is slightly smaller than the stag. In winter the coat is greyish-brown and the white rump prominent. Hock glands stand out as raised cushions on the hind legs.

White rump

Rounded ears

Sika hinds do not have antlers. They bear one calf at a time. When not following its mother, a calf will rest and ruminate, hidden in long grass or undergrowth.

Sika deer *Cervus nippon*

Unlike its close relative the red deer, the sika deer is not native to Britain but was introduced about 120 years ago to a number of deer parks. It originates from parts of eastern Asia, including the USSR, China and Japan, where it is known as the spotted deer. Feral herds in various parts of Britain are derived from escaped or released animals. One of the most flourishing populations is in Dorset, the descendants of deer introduced to Brownsea Island in 1896, some of which swam across Poole Harbour to the mainland. Because of their close relationship, sika and red deer will interbreed and produce fertile hybrid offspring. It is feared that where the two live in the same area, such interbreeding could destroy the purity of the native red deer as a species.

Sika deer are most active at dawn and dusk, when they leave the cover of the undergrowth to graze. Rutting takes place from late September to early November. Mature males (stags) mark out their territories by thrashing bushes and fraying tree bark with their antlers, and fight off rivals to gather a harem of females (hinds). A sika stag's call at rutting time is quite distinctive – a loud whistle repeated several times. Other rutting calls range from a 'raspberry' to a roar.

Distribution patchy. Herds in wild are all derived from escaped or released animals.

The sika deer is spotted in summer, with a bright chestnut-brown coat. Its ears are rounded, and lighter hair on the forehead darkens at the brow to give the deer a frowning look. It has a white tail, often with a dark stripe, and a white rump with dark edging. Stag 32 in. (80 cm) high at shoulder.

91

A rutting bull challenges rivals with roars, and threatens with antlers and hooves. The antlers have flattened main branches and points, and forward-pointing branches with secondary branches.

Short grasses, sedges and lichens are the reindeer's main food. In winter they scrape away snow to expose lichens, perhaps found by smell. In spring they enjoy eating willow and birch shoots.

Calves begin to develop antler pedicles (stalks) at about two months old. The antlers are fully grown before winter ends.

Bulls have larger antlers than cows. Old bulls cast their antlers soon after the September–October rut and grow new ones during winter. Young bulls keep their antlers for a few months after the rut.

Calves are born in May or June, and unlike most other deer, have no spots. They can walk within an hour.

A reindeer's broad, cloven hooves are splayed to spread its weight and prevent it sinking too deep when walking on snow. The hooves make a clicking sound as it walks.

Long, sweeping antlers

Large feet

Reindeer are well able to withstand the icy mountain temperatures in winter. The long winter coat protects a soft, dense underfur that insulates against the cold.

Small herd in Cairngorms near Aviemore, Scotland, introduced in 1952.

The reindeer's long, sweeping antlers and large feet are distinctive. The coat colour varies widely, but many animals are greyish or brownish. The thick winter coat is paler. In late summer bulls grow a prominent mane of white hair that persists through the winter. Bull about 48 in. (120 cm) high at shoulder; cow smaller.

Reindeer *Rangifer tarandus*

About 200,000 years ago reindeer were numerous in Britain, and the icy landscape must have resounded to the noise of thousands of clicking hooves as herds migrated between summer and winter feeding grounds. The herds dwindled, perhaps as the climate changed, and the last survivors lived in Scotland. No one knows for certain when they finally died out. Reindeer were re-introduced to Scotland in 1952, when a domesticated Swedish herd from Lapland was released in the Cairngorm mountains near Aviemore.

In summer, male reindeer (bulls) are usually solitary, joining the herds of females (cows) and young animals for the September–October rut (mating season). After the rut, the bulls separate from the herd but follow it. Cows grow antlers, the only female deer to do so. Although mature bulls shed their antlers in autumn, the cows carry theirs until spring. In winter, therefore, the cows can use their antlers to defend feeding patches cleared for themselves and their calves. Reindeer use their antlers as well as their hooves to scrape the snow off food plants, and in North America are known as caribou, a name derived from a Red Indian word meaning shoveller.

The kids are spotted with white. The spots gradually fade and after two months have disappeared.

When alarmed, roe deer fluff out the pale hairs on the rump, which looks like a large powder puff.

Before mating, the buck sniffs and chases the doe, often in circles. This creates a track known as a roe ring.

Buck

Doe

Roe deer have very short tails. The rump patch is white in winter. The doe also has a long tuft of hairs hanging on the rump.

The grey-brown winter coat that grows in September and October is moulted in the following spring. The deer look scruffy while the winter coat is moulting. A wary deer will often stamp a foot.

Does do not have antlers. A doe may lie down to suckle a new-born kid. For the first few days of its life a kid stays hidden among vegetation, the doe visiting it at intervals.

Black nose

White chin
patch

Buff
rump

Antlers are cast in
November or December.
During winter new ones
grow, protected from frost
by a woolly skin (velvet)
rubbed off by May.

A roe buck disturbed in a woodland clearing is agile
enough to leap gracefully away and clear a fence or
other barrier in one bound as it takes to cover.

A forest deer spreading in
many areas, especially to new
plantations.

A roe deer's summer coat is a sleek, foxy red
with a buff patch on the rump. Its white chin
patch and black nose, sometimes with a white
rim or patch above, are distinctive. The ears
are large and furry inside. A buck's antlers are
roughened (pearled) near the base. About
25 in. (64 cm) high at the top of the shoulder.

Roe deer *Capreolus capreolus*

In the Middle Ages the roe deer, the smallest of our native deer,
was widespread in Britain. Later it gradually disappeared, no
one is sure why, surviving in only a few places. About 100 years
ago the deer was re-introduced to parts of England and has
spread to many woodland and upland areas.

Roe deer generally keep to cover, and are usually seen in
small groups or singly. They are most active at dawn and dusk
and feed mainly on tree shoots or shrubs. Rutting, or mating,
takes place in July and August. A male (buck) establishes his
territory at the end of May, when his antlers are hard and fully
grown. He rubs against trees and bushes to scent them and barks
at rival bucks and chases them off, sometimes fighting. Females
(does) entering the territory are courted and mated, yearling
does being ready for mating before older does. Roe deer are the
only hoofed mammals in which implantation of the fertilised
egg is delayed. It does not occur until December and the young
are born in the following May or June. Twins are common and
there are sometimes triplets. New-born young (kids) seen lying
among bracken, brambles or grass should not be disturbed. The
doe is near by and returns to suckle them several times a day.

95

Deer in the forest

Roe deer are particularly numerous in the coniferous forests of northern Britain, but are shy creatures that will disappear into the trees if disturbed. They are most likely to be seen as they cross a forest ride. At about two years old a roe buck acquires a territory that he normally occupies for life. The size varies, and may cover about 12–75 acres (5–30 hectares). During spring and summer, the buck patrols the boundary of his territory, marking it by rubbing scent from his head glands against trees and fighting with intruding bucks. Roe does also have territories, but do not defend them. They often overlap with those of other does and several bucks.

Deer of any species may cause considerable damage to trees by fraying bark when they clean their antlers, by breaking branches when they thrash with their antlers at rutting time and also by stripping bark to eat. A tree's growth may be distorted, and if bark is rubbed or stripped off all round, the tree will die. Deer also browse on accessible shoots, large deer reaching as high as 6–7 ft (1·8–2·1 m) when standing on their hind legs. In commercial forests the damage can be serious and new plantations in traditional red deer areas are generally fenced off.

As deer have few natural predators, numbers are controlled by selective shooting. High platforms enable deer stalkers to observe deer in the rides and shoot from cover.

Deer prefer to browse on deciduous trees but will eat the tips of conifer shoots. Roe deer do not usually stand on their hind legs when browsing.

In April roe deer clean the velvet from their fully grown antlers by rubbing against trees. Branches may be broken and the bark rubbed off. Red deer do the same thing in August.

All deer like to browse on the brambles that grow in clearings. Roe deer are especially fond of them.

Creatures of habit, deer regularly take the same route through the forest, making well-trodden paths.

A roe buck disturbed as it browses will bound into the cover of the trees. It may clear about 16 ft (5 m) in one leap.

Fallow deer come to graze in the grassy forest rides – the broad paths between stands of trees. Fallow of various colours, muntjac, roe and red deer may all be found in one forest.

Muntjac fawns are born singly at any time of the year. Their spotted coats make them hard to see in the undergrowth. By the time they are eight weeks old the spots have faded.

A buck's antlers are cast in May or June and grow again during summer. They are seldom longer than 4 in. (10 cm). A muntjac has large glands below the eye, in pits almost as big as the eye socket.

When alarmed, a muntjac lifts up its tail to show the white underside – maybe to warn others as it runs away.

A muntjac's tongue is extremely mobile and so long it can reach to lick the corner of its eye. A buck has a V-shaped ridge on the forehead, extending down from his antlers and marked by dark stripes.

Mutual grooming is common between adult bucks and does. A doe lacks antlers and has a black triangular patch on her forehead.

Small and secretive, muntjac are difficult to see in their woodland home. They prefer places where there is plenty of dense undergrowth to give good cover.

Muntjac *Muntiacus reevesi*

Although it is called the barking deer in its native Asia, the muntjac is not the only barking deer in Britain, for several other species also bark. The Chinese, or Reeves's, muntjac was introduced to the Duke of Bedford's Woburn estate in Bedfordshire about 1900. Since then the descendants of escaped or released animals have become well established in England, and are still spreading. Their increase is no doubt due to the fact that females (does) can conceive a few days after fawning and may give birth every seven months. Unlike native British deer, they have no fixed breeding season.

Muntjac are active by day or night but are most often seen at dusk, feeding on grass, brambles and other plants, including ivy and yew. Their winter coats are duller and thicker than their summer coats. In common with all deer, muntjac have scent glands with secretions that are probably a means of communication. They include forehead glands. To mark territory they rub their heads against the ground or a tree to leave scent. A male (buck) establishes a territory that includes the home area of several does and will fight rival bucks. He uses his fine-pointed fang-like teeth as weapons rather than his antlers.

Rounded back

Short antlers

The smallest British deer, the muntjac has a glossy, red-brown summer coat and is distinguished by its rather rounded back. There is some white round the edge of the tail, but the amount varies. The buck has short antlers and prominent fang-like teeth in his upper jaw. Buck about 19 in. (48 cm) high at top of shoulder; doe slightly smaller.

Found in woodland and scrub. Introduced to Bedfordshire, still rapidly spreading.

The tail is short and held close to the rump. It is inconspicuous, especially in winter when the deer's coat is pale grey-brown and very thick.

The deer feed mainly on grasses, rarely browsing on shrubs or trees. Does are slightly smaller than bucks.

Fawns are born about June and are spotted. Twins and triplets are usual, but sextuplets may occur. Many fawns die soon after birth.

As with other ruminants, a period of grazing is followed by a period of rest, during which the deer chews the cud.

Tusk-like teeth

The deer's eyes and nose show up like three black buttons on its alert face. By early summer it has lost most of its thick, grey-brown winter coat.

Chinese water deer *Hydropotes inermis*

The reed swamps and grassy river valleys of north-east China are the original home of the Chinese water deer. It was introduced to England at Woburn Park, Bedfordshire, early this century. Animals bred there were supplied to Whipsnade Park and private landowners, who in turn sent some to parks in several counties. Escapes occurred but few feral populations have become well established. Some deer have flourished in the Cambridgeshire fens and the Broads of East Anglia, where there are wetlands with dense reed beds and clumps of alder providing thick cover. Here they easily escape detection, but sometimes venture onto nearby arable fields.

Feral water deer are usually solitary or in very small groups. In contrast, deer in parks such as Woburn gather in large groups and are easily seen. They are particularly noticeable at Whipsnade, where many range freely through the park or run over the Downs. They leap and bound through the grass with hind legs flung high in the manner of hares. Large numbers often gather on the surrounding fields, especially during November and December, which is the rutting season. They bark, whicker and squeak as they chase one another.

Feral groups in and around Bedfordshire, also Broads, Fens. Park herds elsewhere.

Unlike any other male deer in Britain, the male Chinese water deer has no antlers but has tusk-like teeth protruding about 2¾ in. (70 mm) below its upper lip. The deer is slightly higher at the haunch than at the shoulder, and in summer is a sleek red-brown. Its large, rounded ears are very furry on the inside. Buck (male) about 24 in. (60 cm) high at shoulder.

101

Three-tined antlers

A stag rubs its antlers against a sapling in order to remove velvet or to mark its territory.

A well-grown axis fawn resembles a fallow deer. It can be distinguished by its white bib.

White bib

Axis stags may cast their antlers in any month. As with most deer, hinds (females) have no antlers.

A forest species in the wild. In Britain seen at Woburn and some other parks and zoos.

The attractive axis deer has a rich brown summer coat heavily dappled with white spots, and it has a white bib. Winter colouring is slightly darker. A stag's three-tined (pointed) antlers can grow up to 30 in. (76 cm). About 37 in. (95 cm) high at shoulder.

Axis deer *Axis axis*

Native to India and Sri Lanka, the spotted axis deer can be seen in Britain in a few zoos and deer parks such as those at Woburn and Whipsnade, both in Bedfordshire. It is the most attractive of India's eight species of deer, and its Hindustani name is *chital*, which means 'spotted deer'. Axis is said to be the name that was given it by the Roman administrator Pliny the Elder in his *Natural History*, written in the 1st century AD.

In its homelands, the axis deer lives mainly in lowland forest areas where there is good cover as well as good grazing. It also ventures from the forest to feed on the crops of nearby villages, and this has led to many deer being killed. Earlier this century herds could be reckoned in their thousands, but now they are much fewer in number.

As with most deer, axis deer graze by day or night, or browse on shrubs. When alarmed, they emit a shrill whistle. There is no fixed breeding season, and young calves may be seen among the herds in Britain's parks at any time of the year. Nor is there a fixed season for stags (males) to shed their antlers, so all stages of antler growth, as well as hard antlers cleaned of velvet, can be seen among animals in the herd at any time.

Backswept
points

Long tail

Calves are born between
March and May. They have
spotted coats and the leggy
look of a foal.

All the Père David's deer now in existence
are descended from the 11th Duke of
Bedford's herd at Woburn.

Strangely built for a deer, Père David's
deer has a squarish muzzle, antlers with
backward-sweeping points, large feet
that click like a reindeer's, and a long
tail. In summer its coat is tawny red
flecked with grey; in winter it is longer
and greyish-buff. About 48 in. (120 cm)
high at top of shoulder.

Père David's deer *Elaphurus davidianus*

No one knows how the Père David's deer, or Mi-Lu, behaves in
its wild state, because by the time it was first described to
zoologists in 1865 it was probably already extinct in the wild. A
French missionary and explorer, Père Armand David, discovered a herd in the walled Imperial Hunting Park near Peking
(Beijing) in China, and some were exported to Europe. This was
fortunate because the Peking herd was killed off – by flooding,
by hungry peasants and during the Boxer war around 1900.

The 11th Duke of Bedford, who had been sent a pair of the
deer from France, became interested in the animals and acquired
all those still alive in Europe. On his Woburn estate in Bedfordshire he began to breed the deer in captivity to save the species
from extinction. They thrived, and from the Woburn herd Père
David's deer have been sent to zoos throughout the world,
and to reserves in their native China. A large deer comparable in
size with a reindeer, Père David's deer has a rounded, horse-like
rump and antlers of unusual shape. It grows new antlers during
the winter, and they are a striking sight in their protective
velvety skin. The stags (males) fight during the rut in July and
August as each tries to keep his harem of hinds (females).

Can be seen in Woburn park,
Bedfordshire, and some other
safari and zoological parks.

During the rut in July and
August, stags wallow in mud,
plastering it on their bodies and
antlers.

A stag will festoon
its antlers with
vegetation during
the rutting season.
Hinds have no antlers.

103

The goats browse on shrubs such as gorse and heather, and will also eat leaves and shoots from trees. They sometimes stand on their hind legs to reach a branch.

Horns grow continually – a ram's can reach 30 in. (76 cm) in length. The goat's long horns help to distinguish it from hill sheep, which have short, coiled horns.

Males spar for supremacy at rutting time. Rutting starts in mid-August in northern Scotland, but further south it is generally mid-October.

The young (kids) are born in January in the north, March or April in the south. One kid is usual, twins are uncommon.

Its shaggy coat distinguishes a feral goat from the modern, short-haired domestic goat. A modern goat may be about twice the size of a feral goat.

Sure-footed and agile, goats can forage on rocky slopes, and may be surprisingly hard to see. Females tend to keep to one place, but males often range over many square miles.

Spreading
horns

Shaggy
coat

Moors, cliff tops and rocky hillsides are the home of the
feral goat. This group belongs to a herd on Skokholm
island off the Dyfed coast, South Wales.

Feral goat *Capra* (domestic)

Stone Age farmers first brought goats to Britain, and there have
been herds of feral goats – descendants of domestic animals gone
wild – in mountainous areas for well over 1,000 years. Two-
hundred years ago, shaggy feral goats were commoner than
sheep in many places, but today the remaining herds are small
and isolated. There is considerable variation in colour and size
between individuals, as with many feral species once the rigours
of selective breeding have been removed. Because isolation
prevents interbreeding, the goats in one place tend to be different from those in another.

For most of the year the goats keep to high, rocky mountain-
sides or cliff tops, in winter descending to grassy valleys or the
neighbourhood of farms in search of food. Towards the end of
winter the herds disperse and females go off to give birth to
kids. At first a kid stays hidden among boulders, visited by its
mother two or three times a day. When about ten days old a kid
is strong enough to follow its mother, and by summer females
and youngsters have gathered into a herd. Kids are too big for
most predators, but many die of exposure. Adults are better able
to withstand damp and drizzle, and often live for five years.

Isolated herds in rocky areas of
the north and west. Small
herds on some islands.

The small, shaggy feral goat is a free-
living descendant of domestic goats of
the past, before modern breeds were
developed. Both sexes have horns,
which may sweep back or, more often,
spread outwards. Male 30 in. (76 cm)
high at shoulder. Female smaller.

The fleece contains both wool and hair, and moults in June. St Kildans plucked the wool by hand – sheep were not shorn.

There are buff markings above each eye, and the pupils of the eyes are rectangular. On the ram's horns, dark annual growth rings can be seen between groups of ornamental ridges.

In late summer the sheep population of St Kilda is at its peak; every few years there are so many that they eat all the available grass.

During the rutting season in late autumn, rams may be seen sniffing with heads and necks outstretched as each one follows a ewe.

The rams establish social dominance at rutting time by behaviour such as blocking – pushing against one another with one foot raised.

Soay lambs are born in mid-April. Twins are not uncommon. Many lambs die at the end of winter when food is short.

Fawn coat

Dark brown coat

A primitive breed found on Soay and Hirta islands in the remote St Kilda group.

Slightly built Soay sheep resemble the primitive domestic sheep of Stone Age Britain. They are mainly dark brown or, less commonly, fawn, with paler underparts and rump. Ewes have small horns or none at all; rams have heavy, coiled horns. Ram 22 in. (55 cm) high at top of shoulder, ewe smaller.

The Vikings probably first established Soay sheep on St Kilda. Isolated for hundreds of years, the breed retained its primitive characteristics.

Soay sheep *Ovis* (domestic)

Tiny, dark brown Soay sheep on the St Kilda islands in the Outer Hebrides are the only completely wild sheep in Britain. They are the kind of sheep kept centuries ago before selective breeding developed more productive animals, and would have died out but for their isolation on remote St Kilda. They supplied the islanders with wool and meat for hundreds of years until the last people left the islands in 1930.

Because the St Kilda sheep are no longer managed in any way, the population tends to build up to over 1,000 animals. Then, as a result of overgrazing, as many as a third of them die of starvation at the end of winter. The population builds up again in the following three or four years to repeat the cycle.

Soay sheep are kept in small flocks in many mainland zoos and country parks, for they are attractive and need little attention. Their fleece, soft and naturally brown, is popular with people who make garments from homespun wool. Because the sheep are not as heavy as ordinary domestic breeds, their hooves do not cut into turf and encourage soil erosion. They are often kept on nature reserves where gentle grazing is needed to maintain short turf rich in orchids and other flowers.

Heather is the wallaby's main food, especially in winter, but it also eats grass and bracken and browses on conifer seedlings and bilberry shoots. It sometimes uses its front feet to grasp plants.

The wallaby can withstand temperatures that are moderately cool, but is badly affected by very cold weather. In deep snow it has difficulty in moving about and finding food.

The young are born singly at any time of year. A tiny new-born wallaby crawls up into its mother's pouch and attaches itself to a teat. It is suckled for several months. Young in the pouch are called joeys.

When it bounds fast, the wallaby's long hind feet and long, thick tail are very conspicuous. It uses only its hind legs, tucking its front legs up against its chest.

Older joeys often leave the pouch to feed. Some are so big they have difficulty in getting back, and their hind feet may be left sticking out. A young wallaby becomes independent from about 10 to 12 months old.

When a joey is a few months old it begins to take solid food, and often grazes from the pouch as its mother bends forward to nibble the grass.

Red-brown
shoulder patch

Black-
tipped
tail

Black-
tipped
feet

Occurs in a small area of the
Peak District. A few in Sussex
have died out.

The shy red-necked wallaby tends to
hide in thick scrub. It takes its name from
the red-brown patch on its nape and
shoulders. Its feet and tail are tipped with
black. Male 24 in. (60 cm) head and
body, 25 in. (64 cm) tail. Female slimmer
than male.

A rare and rather surprising sight in the British countryside, the red-necked wallaby is usually seen on its
own or occasionally in small groups.

Red-necked wallaby *Macropus rufogriseus*

Kangaroo-like red-necked wallabies from the scrublands of
Tasmania must be the most surprising creatures to be found
living free in Britain's countryside. Many have escaped from
zoos or wildlife parks in the past 150 years, but only a few have
survived to breed in the wild. The best-known British population, just a few dozen, lives in part of the Peak District National
Park. They are the descendants of wallabies that escaped from the
grounds of a local mansion in about 1940. Another small
population (now extinct) became established in Ashdown Forest
in Sussex. As wallabies are still kept in many zoos and wildlife
parks, further escapes are likely. There have been reports of some
seen in the Loch Lomond area.

Britain's few wild wallabies are rather shy and rarely seen.
They are well suited to living in thickets and woods, and are used
to the cool, wet climate of Tasmania, but harsh winters and
deep snow cause many to die. The wallaby's chief hazards are
probably disturbance by people and competition for food from
large numbers of sheep. Roads and electrified railway tracks, on
which some have been killed, also prevent wallabies from
spreading to better habitats, as do wire fences.

The female builds a breeding nest of grass and leaves. One litter of three or four young is born in April or May. They become independent when about a month old.

As moles often live in low-lying areas their burrows are likely to become flooded. But moles are able to swim to safety, paddling with all four limbs.

The mole is not blind, but each eye is only about the size of a pinhead. It finds its way in the darkness underground by means of sensitive whiskers and touch sensors on its nose.

The mole is able to run backwards through its tunnels. Sensitive whiskers on its tail detect any obstacles in its path, and its velvety fur will lie backwards or forwards so that it does not jam against the tunnel walls.

Colour variation is more frequent than among other animals, perhaps because the variant moles, living underground, are less likely to attract the attention of predators. Apricot, creamy-white, pale grey and piebald moles may be found.

Upright tail

Silky black fur

The mole is built for tunnelling, with big, heavily clawed front feet and a solid, muscular body covered with silky black fur. It has a long, tapering nose and a short tail, which is carried upright. Male 6 in. (15 cm) head and body, female slightly smaller.

Heavily clawed front feet

Found in all types of country except high moors, mountains. Lives mostly underground.

Moles are most abundant in pasture and grassland. Their fur is water repellent, and squeezing among roots and through tunnels keeps it combed and sleek.

Mole *Talpa europaea*

Much of a mole's life is spent in a burrow system that it tunnels underground, so it is not often seen. Its presence is evident from the soil heaps (molehills) it makes while tunnelling, and in medieval times one of the names given to the mole was *moldewarp* (earth thrower). Active throughout the year, moles are found in most places except ground above 3,000 ft (roughly 1,000 m) and in very acid soils. In woodlands, where they are quite common, molehills are often hidden by leaves.

Moles come to the surface to collect nesting material – dry grass and leaves – and also to look for food when the soil is dry. Their main food is earthworms, which, being full of soil, are heavy but not very nutritious, so moles eat at least half their body weight in food every day. Young moles probably come to the surface to find new homes when they leave their mother's burrow. Moles are most vulnerable to predators when above ground, even though they emerge mainly at night. Their skin glands make them distasteful to carnivorous mammals, but large numbers are eaten by tawny owls and barn owls. About a third of the mole population survives for more than a year, but not many live to be three years old.

111

A mole regularly patrols its tunnels to eat whatever soil animals – such as worms, beetle larvae and slugs – have fallen from the walls. Tunnels are longest and molehills most numerous in poor soil, where the mole's food supply is sparse.

Although it can travel backwards in its tunnel, a mole can also turn in the opposite direction by doing a forward roll.

How moles live and feed underground

Because of their digging activities, moles are considered a nuisance by many gardeners and farmers. Most tunnels are quite near the surface, and tunnelling may interfere with the roots of garden plants and crops. Occasionally tunnels are so near the surface that the soil is forced up in a long ridge – once these were thought (wrongly) to be 'love runs' made by male moles seeking a mate. Molehills each contain about 2 lb (1 kg) of soil. They disfigure lawns and in fields are an inconvenience to farm machinery such as combine harvesters; they may even damage them. But moles can also be quite useful. They eat a lot of insect larvae that are damaging to grass and root crops, and their tunnelling helps to aerate the soil – important in peaty and waterlogged areas.

In days gone by, country parishes often employed a professional mole-catcher to trap and dig up moles in places where they were not wanted. In one season, a mole-catcher might catch 1,000 moles and earn a small income by selling the skins for hat and coat trimmings. As late as the 1950s about a million moles were being trapped every year in Britain. Since synthetic furs became common, however, catching moles for their skins has become no longer worth the effort.

When tunnelling, a mole uses one front foot to force soil upwards into a molehill while it braces the other, and its hind feet, firmly against the walls of the tunnel. It can move twice its own weight of soil – some 8 oz (over 200 g) – a minute.

Each mole has its own burrow, a system of firm-walled tunnels about 2 in. (50 mm) wide and 1½ in. (40 mm) high which may be about 200 yds (roughly 200 m) long.

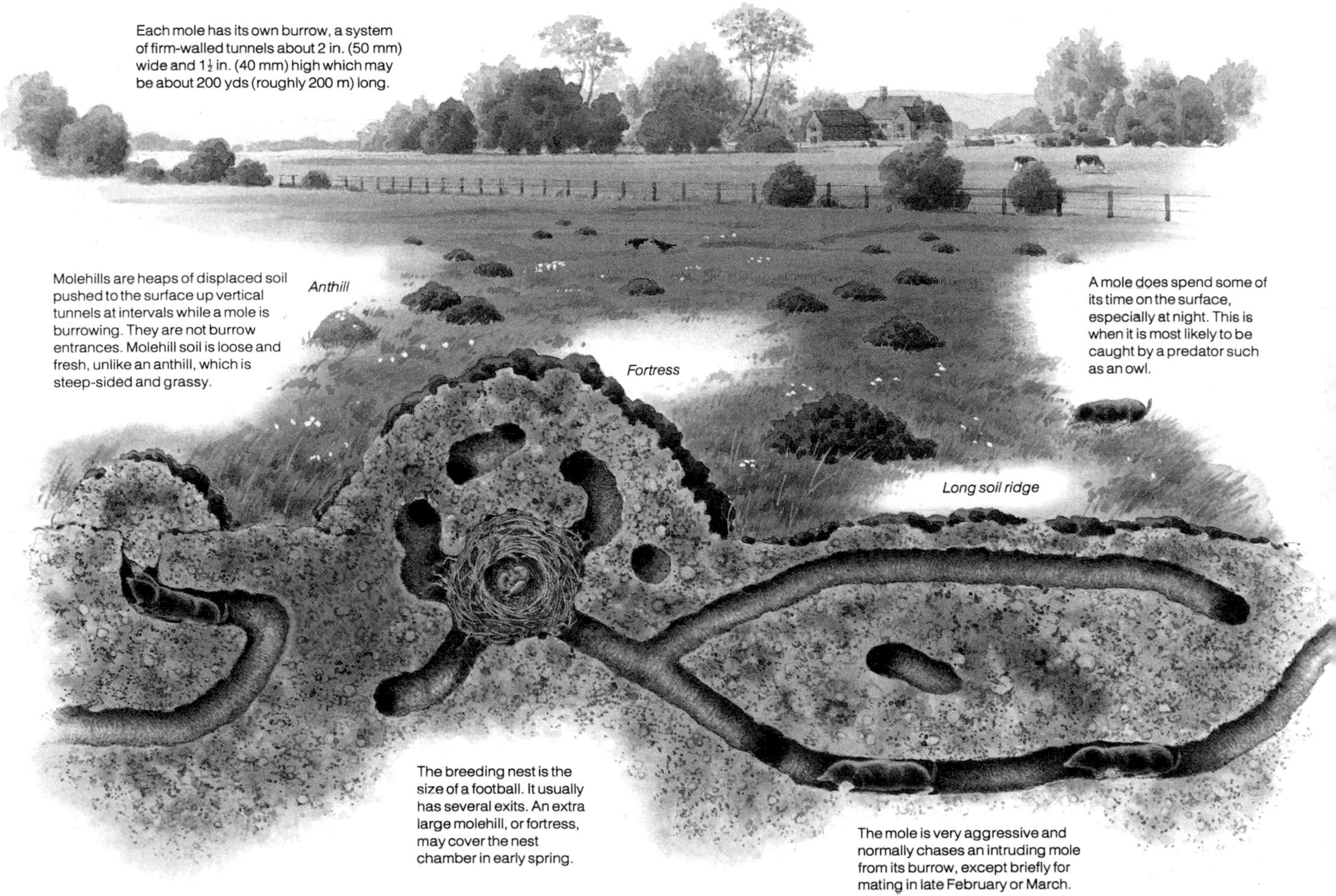

Molehills are heaps of displaced soil pushed to the surface up vertical tunnels at intervals while a mole is burrowing. They are not burrow entrances. Molehill soil is loose and fresh, unlike an anthill, which is steep-sided and grassy.

Anthill

Fortress

A mole does spend some of its time on the surface, especially at night. This is when it is most likely to be caught by a predator such as an owl.

Long soil ridge

The breeding nest is the size of a football. It usually has several exits. An extra large molehill, or fortress, may cover the nest chamber in early spring.

The mole is very aggressive and normally chases an intruding mole from its burrow, except briefly for mating in late February or March.

113

Voles that feed on contaminated roadside plants have unusually high lead levels in their bodies. It is not known whether this does them any serious harm.

A newly built road may cut across an established badger trail, and badgers using it may be run over. Where such trails occur, many roads now have built-in badger subways.

Moles living on motorway verges are safe from farm ploughs as well as from mole-catchers. Molehills are often numerous on the central reservation, and as they do not contain road rubble, the moles must have crossed the road above ground, not by tunnelling.

Voles, mice and beetles on mown verges are prey for hovering kestrels.

Deer often dash across roads at dawn or dusk. Reflectors or mirrors are sometimes fitted to fences or trees at bends to warn deer of approaching vehicles.

Rabbits dig their warrens in well-drained, undisturbed embankments. They usually feed in nearby fields, not on oily, gritty, roadside vegetation.

Early in the morning crows feed on the night's road casualties, such as hares, hedgehogs and rabbits.

Motor roads as highways and havens

The amount of grassland and scrub along the broad verges of Britain's major roads and motorways exceeds the total area of all the country's nature reserves put together. People rarely walk on roadside verges (it is in fact forbidden beside motorways), so they offer a comparatively safe home to animals that can tolerate traffic noise. Moles, particularly, benefit because no one needs to plough up their burrows, and because traffic vibration brings worms, their major food, to the surface. Small mammals such as voles and shrews abound, and foxes and rabbits are quite common. The varied plant life offers food as well as cover to some small mammals, but it is often contaminated with salt and oil from the road surface and with lead from car exhaust fumes.

Road verges serve as corridors along which animals can spread into cities and across wide areas of country. The actual crossing of roads is a problem, however. Many mammals do manage it – even slow-moving moles – but rabbits, hedgehogs, deer and other animals active at night are often dazzled by vehicle headlights and run over. Many thousands are killed on the roads every year; their corpses serve as a food supply for scavengers such as foxes, crows and magpies.

From three to five young are born, usually in June or July, in a nest of leaves and grass. The young are blind and pink at first, but sprout a few white, bristly spines within hours of birth.

An adult hedgehog has some 5,000 spines on its back. When it rolls into a ball, its spiky coat protects it from all but the most determined predators. The young, which have fewer spines and weaker rolling-up muscles, are more vulnerable.

During the first two weeks of life, young hedgehogs grow more and more brown spines alongside the white spines. Their eyes open at about 14 days old. By the time they are from five to six weeks old, they have more than 2,000 spines.

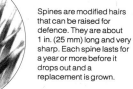

Spines are modified hairs that can be raised for defence. They are about 1 in. (25 mm) long and very sharp. Each spine lasts for a year or more before it drops out and a replacement is grown.

Spiny coat

Hairy face

White hedgehogs are not uncommon. Light colouring may attract predators, but protective spines offset the disadvantage, so the variation still occurs.

Hedgehog *Erinaceus europaeus*

Britain's only spiny mammal, the hedgehog has been a familiar creature of gardens, hedgerows and meadows for centuries. It features in some curious folk tales. They tell of hedgehogs picking up fruit on their spines and of hedgehogs sucking milk from cows, but such behaviour is improbable.

Beetles, caterpillars and earthworms make up most of the hedgehog's diet, but it will also eat birds' eggs, slugs, snails and carrion. In winter, when such foods are in short supply, the hedgehog hibernates in a sheltered nest of leaves and grass and lives off reserves of body fat built up in the autumn. Hedgehogs are ready to breed in April, soon after hibernation ends. Most young are born in early summer, but late litters in September are common, especially if the first litter has been lost. Late-born young often do not survive the winter. A mother disturbed in her nest may eat or abandon new-born young, but will carry older ones by the scruff of the neck to a safer place. Youngsters start to take solid food at about a month old, and at night the mother leads her family out in procession to forage. The young become independent at about six weeks old. The male takes no part in rearing them, and may not even live near by.

Widespread on farmland and in urban areas. Scarce on moors, in conifer forests.

Unmistakable because of its spiny coat, a hedgehog also has long, coarse hair on its face and underparts. It is active mostly at night, but young or sickly animals may be seen by day. It relies mainly on smell to find food. About 10 in. (25 cm) long.

117

Although hedgehogs can swim well, they often drown in swimming pools and even tiny garden ponds because they cannot climb the smooth sides. A strip of wire netting fixed to the side aids escape.

Cattle grids, trenches and similar holes are a danger to hedgehogs because they often fall in and cannot get out. Their spines may act as a cushion and make them less afraid of falling than most animals. Many cattle grids now have ramps or tunnels which allow hedgehogs to escape.

Helping hedgehogs to survive

Although few animals prey on them, hedgehogs face many hazards to survival. Every year thousands are killed by motor vehicles because they react to danger by rolling up. There is little evidence to support the theory that more are tending to run from danger, nor would running necessarily save their lives, although they can run quite fast. As hedgehogs feed on garden pests, it is in a gardener's interest to encourage them. Accidents that befall some hedgehogs in gardens are drowning in ponds or becoming entangled in netting. The use of garden chemicals such as insecticides and possibly slug pellets is also a threat to hedgehog survival. Many such chemicals are present in minute quantities in beetles and caterpillars, and as hedgehogs eat hundreds of them every month they can soon accumulate enough poison to damage their health.

Hedgehogs face their greatest risks during hibernation. They may die of cold or be disturbed by fire or flood, or someone may wreck their nest. Gardeners can help by leaving piles of fallen leaves undisturbed behind sheds and under hedges. More than half of all hedgehogs do not survive their first winter. The remainder may live two or three years, rarely five or six.

Hedgehogs are agile and active animals. They will climb rough stone walls and even some garden fences, using their strong claws and surprisingly long legs.

Extra food such as bread and dog food with goat's milk, or cow's milk well diluted with water, helps garden hedgehogs to fatten up for winter hibernation. They will go more than 440 yds (400 m) for such a treat. A hedgehog will die during winter if its reserves of body fat are inadequate.

If a hedgehog is seen to froth at the mouth and twist itself about, it is not sick but is spreading the frothy saliva on its fur and spines with its tongue. The purpose of this behaviour is unknown, but it is quite normal.

Hedgehogs sniff through garden litter to nose out woodlice and slugs. Eating poisoned creatures can lead to a hedgehog's death. Use garden chemicals sparingly.

119

The carrion crow finds the hedgerow a good place to scavenge for dead rabbits or voles. Hawthorn berries provide sustenance for noisy flocks of fieldfares.

Winter in the hedgerow

On bitterly cold winter days the countryside is still and silent, with only a few birds or an occasional rabbit abroad. Winter weather makes it difficult for animals to keep warm, and frost and snow greatly complicate the business of finding food. Digging for worms is impossible in frozen ground, and it is too cold for many small creatures such as beetles to be found.

Hedgehogs and dormice avoid the problem by hibernating in a weatherproof nest, often at the bottom of a hedge. Instead of trying to find enough food to provide sufficient energy to keep warm, they allow their bodies to cool down and remain inactive for weeks on end and until better weather returns. For some animals, snow is a welcome protection. On a clear, frosty night, a brown hare crouched on the surface may have to endure an air temperature of 18°F (−8°C). But a mouse under the snow is protected from the cold air and exposed only to snow temperature – about 32°F (0°C). Inside its hedgerow nest or burrow it can even generate enough heat to be quite warm – perhaps at a temperature of 68°F (20°C) or more.

Rabbits scrape away the snow with their forepaws to search for grass and other plants to eat. Hedgerows often shelter rabbit warrens.

Unlike the hedgehog, the field vole is active throughout the winter. It tunnels among the grass underneath the snow, eating blades and stems.

A grey squirrel's winter drey in a hedgerow oak is easy to see when the tree is bare. The squirrel does not hibernate and is often out in the snow searching for buried acorns.

A magpie's domed nest is very similar to a grey squirrel's drey, but it does not have any leaves still clinging to the outside.

Old summer nest of harvest mouse

Winter nest of harvest mouse

Inside its winter nest, a harvest mouse is warm and snug under the snow layer. It does not hibernate, but forsakes the upper areas of tall grass and shrubs to live on the ground.

Dead leaves make the most weatherproof winter nest for a hibernating hedgehog. The hedgehog carries the leaves to its chosen site and piles them in a heap supported by brambles or brushwood. It burrows into the heap and shuffles round until the leaves are all firmly packed into the walls of the nest.

IDENTIFYING RATS, MICE, VOLES, SHREWS AND DORMICE

When trying to identify small mammals such as rats, mice, voles and shrews, look at the shape of the nose, the size of the ears and the length of the tail in relation to the head and body. These are often crucial identification points. Colouring and size alone can be deceptive, especially with young animals; a young brown rat, for example, can be mistaken for a mouse. Also, there are brown versions of the black rat and black versions of the brown rat, but they can be distinguished by their ears and tail length.

Mice generally have pointed faces, big eyes, prominent ears and a long, thin tail that is usually longer than the head and body. Rats are like large, heavily built mice with coarse, shaggy fur and scaly tails. Voles are chubby and round-faced with short noses and fairly small eyes and ears. They have long, silky hair and their tails are comparatively short – about half the length of head and body, or less. Shrews are very tiny and have very long, narrow, pointed noses, small ears and pin-head size eyes. Their tails are usually quite hairy and vary from half to three-quarters of head and body length. Whereas voles and several species of mice have fur that is black at the base – visible when blown back – shrews and dormice have fur that is all one colour.

Orange-coloured flanks

Yellow mark on chest

Broad yellow chest-band

Yellow-necked mouse
Apodemus flavicollis
Head and body 4 in. (10 cm).
Uncommon. Heavier than
wood mouse. Page 145

Wood mouse
Apodemus sylvaticus
Head and body 3¾ in.
(95 mm). Page 141

Scaly tail

House mouse
Mus musculus
Head and body 3¼ in.
(83 mm). Has greasy fur and
strong smell. Page 139

Juvenile wood mouse
Distinguished from house
mouse by larger ears, longer
hind feet and longer, thinner,
hairy tail.

Orange rump (adult)

Blunt nose

Harvest mouse
Micromys minutus
Head and body 2½ in.
(64 mm). Uncommon. Tail
used for grasping. Page 137

Finely furred ears

Tail shorter than head and body

Brown rat
Rattus norvegicus
Head and body up to 11 in.
(28 cm). Fur occasionally
black. Page 133

Young brown rat
Distinguished from mouse by
feet, which are larger and
heavier in relation to body,
and by thicker tail.

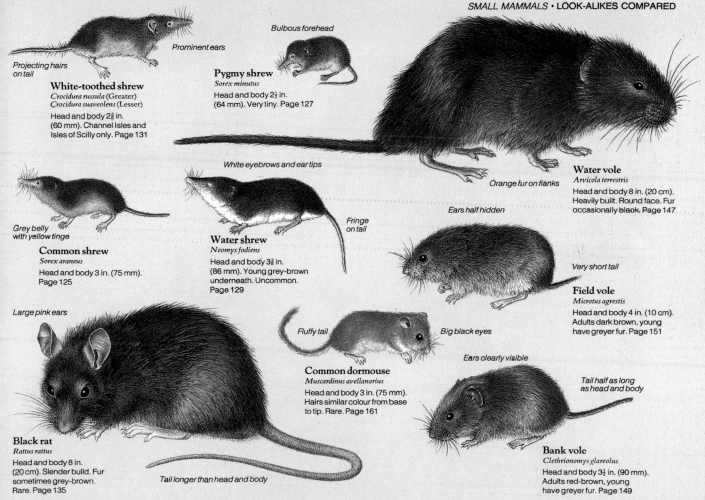

Projecting hairs on tail

Prominent ears

Bulbous forehead

White-toothed shrew
Crocidura russula (Greater)
Crocidura suaveolens (Lesser)
Head and body 2⅜ in.
(60 mm). Channel Isles and
Isles of Scilly only. Page 131

Pygmy shrew
Sorex minutus
Head and body 2½ in.
(64 mm). Very tiny. Page 127

White eyebrows and ear tips

Orange fur on flanks

Water vole
Arvicola terrestris
Head and body 8 in. (20 cm).
Heavily built. Round face. Fur
occasionally black. Page 147

Grey belly with yellow tinge

Fringe on tail

Ears half hidden

Common shrew
Sorex araneus
Head and body 3 in. (75 mm).
Page 125

Water shrew
Neomys fodiens
Head and body 3⅜ in.
(86 mm). Young grey-brown
underneath. Uncommon.
Page 129

Very short tail

Field vole
Microtus agrestis
Head and body 4 in. (10 cm).
Adults dark brown, young
have greyer fur. Page 151

Large pink ears

Fluffy tail

Big black eyes

Ears clearly visible

Tail half as long
as head and body

Common dormouse
Muscardinus avellanarius
Head and body 3 in. (75 mm).
Hairs similar colour from base
to tip. Rare. Page 161

Black rat
Rattus rattus
Head and body 8 in.
(20 cm). Slender build. Fur
sometimes grey-brown.
Rare. Page 135

Tail longer than head and body

Bank vole
Clethrionomys glareolus
Head and body 3½ in. (90 mm).
Adults red-brown, young
have greyer fur. Page 149

Loud squeaking usually betrays the presence of the belligerent common shrew as it fights to defend its territory against a fellow shrew. It cannot bear another's presence, except when mating.

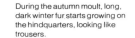

During the autumn moult, long, dark winter fur starts growing on the hindquarters, looking like trousers.

In spring, it moults its winter coat to grow shorter, lighter summer fur that starts to appear on the head.

A female may have a patch of white hairs on the back of her neck in summer, the hair roots being injured when the male seizes her with his teeth while mating.

Earthworms are one of the common shrew's main foods – a lack of them may limit its numbers in places such as moorland. It also eats other soil animals and many insects, consuming almost its own weight in food daily.

Small
rounded ears

Long
pointed
nose

The common shrew lives mostly underground and has poor eyesight. It relies mainly on its keen sense of smell and its long whiskers to find its way about.

Common shrew *Sorex araneus*

No common shrew can tolerate another in its territory, except briefly at breeding time. When two common shrews meet, both scream at each other with shrill, aggressive squeaks, which is perhaps why the term 'shrewish' is applied to a scolding woman. The squeaking can be heard for some distance and is often the only sign that shrews are near by. They are rarely seen because they spend three-quarters of their time underground, but are probably one of Britain's most abundant mammals.

Active by day or night, the shrew is constantly on the move, twittering and muttering and poking its long nose here and there as it scurries along in search of food such as woodlice. It snatches a brief rest every hour or two, but expends so much energy in its bustling life that it will starve to death if it goes without food for more than about three hours. The shrew forages in the soil or leaf litter, along tiny tunnels it digs itself or which have been made by small rodents.

Common shrews may live to be a year old – few live longer. Many fall prey to owls, other predators being deterred by the foul-tasting glands in the shrew's skin, which also give it a distinctive odour. Cats may kill shrews, but rarely eat them.

Smaller than a house mouse, the common shrew has a long, pointed nose, tiny eyes, and small, rounded ears set close to its head. A female may have several litters of six or seven young a year, from April onwards; they become independent at one month old. About 3 in. (76 mm) head and body, 1½ in. (40 mm) tail.

Common in hedgerows, fields and woods, but scarce on moorland. None in Ireland.

Barn owls often pounce on pygmy shrews, guided by the rustle of the shrew's progress. Owls do not mind the foul taste of the shrew's skin glands.

Pygmy shrews often share the same habitats as common shrews, but usually take care to avoid a conflict with their larger, more aggressive relatives.

Breeding takes place mainly from April to August. Most young are born in June or July, in a nest underground. There are from four to seven young in a litter.

Alongside a ½ in. (13 mm) yew berry, the pygmy shrew's very small size is evident. Even when fully grown, it weighs less than a 10p coin.

The pygmy shrew is small but hardy and can thrive on open moorland, even on the cold heights of Ben Nevis, Britain's highest mountain.

Narrow,
pointed
nose

Thick tail

The tiny pygmy shrew is almost uniform brown in colour, with a narrow, pointed nose, small eyes and a thick tail more than half **as** long **as** its head and body. About 2½ in. (64 mm) head and body, 1½ in. (40 mm) tail.

Woodlice are common in the pygmy shrew's diet. It eats Its own weight in food every day, amounting to only a fraction of an ounce (about 6 grams).

Found on farmland, moors and in forests. Widespread, but scarcer than common shrew.

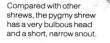

Compared with other shrews, the pygmy shrew has a very bulbous head and a short, narrow snout.

Pygmy shrew *Sorex minutus*

The pygmy shrew is Britain's smallest mammal, not much bigger than a stag beetle or longhorn beetle. It is so tiny that it is near the limit at which a warm–blooded animal can exist – if it were any smaller its body surface would be too extensive for its bulk and it would lose heat too rapidly to maintain a warm body temperature. The shrew does in fact lose so much energy as body heat that it must constantly search for food, and will starve to death if it fails to eat for more than two hours. Hardly ever pausing for more than a few minutes in its busy foraging, the pygmy shrew may explore more than 1,500 sq. yds (about 1,250 sq. m) of territory regularly. It bustles along shallow tunnels made by other animals in the soil, leaf litter or vegetation, and avidly consumes small soil creatures such as spiders, tiny beetles and insect larvae.

Females give birth to several litters each summer. The youngsters grow quickly and leave the nest at about three weeks old. Those born in early summer may be raising families themselves within a few weeks. But many shrews die within a few months of birth, and those that survive to breed the following year do not live through a second winter.

127

All-black animals are found more often than among other types of shrew. Other colour variants are rare.

A keel of bristly hairs along the underside of the tail and a hairy fringe on each hind toe are aids to paddling and steering while swimming.

Prey is usually caught from behind and, if taken in the water, hauled ashore to be eaten. Mild poison in its saliva may help the water shrew to subdue large prey such as a frog.

Young water shrews leave the nest at about four weeks old. Families often travel in procession; the mother leads and the others hold the one in front.

Black coat

When it dives, the water shrew has to paddle fast because air trapped under its fur makes it buoyant. The trapped air gives it a silvery look under water.

White eyebrows and ear tips

Water shrew *Neomys fodiens*

All shrews tend to be more abundant in damp places, where their prey is common – worms, insect larvae and small spiders, for example. Water shrews will also take to the water to hunt in ponds and streams for other prey such as fish, tadpoles and even quite large frogs. However, they also often live away from the water, even on dry downland, and some inhabit stony beaches where they probably feed on sandhoppers and flies along the high-tide line. Active by day or night, the shrews eat roughly their own weight in food daily.

Except for females raising families, water shrews normally live alone in shallow burrow systems they dig themselves. Sometimes a burrow is in a bank and has an underwater entrance. Two or three litters containing from three to eight young may be born in a year, from May onwards. Some females have their first litter when only two or three months old, but most do not breed until the summer following their birth. Although they may live for up to 18 months, water shrews mostly die young or are taken by predators such as owls, pike and mink. Numbers tend to fluctuate, the shrews apparently disappearing from an area then reappearing after several years.

Found in farmland, woods and hedgerows. Widespread, but may be scarce in some areas.

The water shrew's black coat and silvery-white underparts meet in a sharp line along its flanks, and its eyebrows and ear tips are often white. Juveniles have duller coats than adults. About 3⅜ in. (86 mm) head and body, 2¼ in. (55 mm) tail.

129

The shrew has big scent glands in the skin of the flanks and by the tail. They give off a musky scent. The flank glands are sometimes exposed when the animal is running about.

Insects and other small creatures such as snails are all part of the shrew's diet. It eats nearly its own weight in food every day.

Greater white-toothed shrew
Crocidura russula

Sometimes called the musk shrew because of its strong scent, the greater white-toothed shrew lives in farmland and hedgerows on the Channel Islands of Alderney, Guernsey and Herm. It is particularly common around buildings in dry places. It differs from the lesser species only in its tooth pattern.

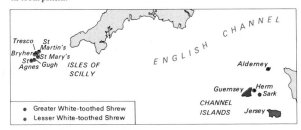

Tresco St Martin's
Bryher St Mary's
St Agnes Gugh
ISLES OF SCILLY

ENGLISH CHANNEL

Alderney

Guernsey Herm
Sark

CHANNEL ISLANDS Jersey

● Greater White-toothed Shrew
● Lesser White-toothed Shrew

In Britain found only in the Isles of Scilly and the Channel Islands. Except on Jersey, white-toothed shrews are the only shrews present. The lesser and greater species never occur together on the same island.

The young are born in a nest of grass and leaves often sited amid boulders or under tree roots. There may be three or four litters of up to six young in one season.

The only shrew found in the Isles of Scilly, the lesser white-toothed shrew needs a lot of food but can go longer without eating than other kinds of shrew.

Projecting tail hairs

The lesser white-toothed shrew, or Scilly shrew, is often found on the shores of the Isles of Scilly, where it forages for sandhoppers among damp seaweed and boulders. Big ears and the projecting hairs on the tail are characteristic of white-toothed shrews. About 2⅜ in. (60 mm) head and body, 1½ in. (40 mm) tail.

Lesser white-toothed shrew *Crocidura suaveolens*

Unlike any other British shrews, which all have red–tipped teeth, white-toothed shrews have all–white teeth. There are two species, lesser and greater, neither of which is found on the mainland. The lesser white-toothed shrew is found on the Isles of Scilly and on Jersey and Sark in the Channel Islands. Its peculiar distribution suggests that it may have been introduced to the islands in boatloads of animal fodder from southern or eastern Europe, where it is also found. The greater white-toothed shrew of the Channel Islands is common in France.

Normally the Scilly shrew lives under logs or dense vegetation or among stones, where it is out of sight of predators such as kestrels. It does not wander far, usually keeping within about 50 yds (50 m) of its nest. The mild climate of the islands permits a long breeding season, stretching from early summer until late autumn. The young take about three weeks to reach weaning age, by which time their mother is often pregnant again. A female sometimes leads her family out in a long procession – known as caravanning – each one grasping the animal in front. Most white-toothed shrews die at an early age, but they can live for well over a year.

131

Often seen in the water, especially near a sewer, a swimming brown rat can be distinguished from a water vole by its pointed nose and greyer fur.

A rat colony may be found among a pile of rubbish in a hedgerow. The rats live in underground burrows with trampled runways leading to them. Young rats first emerge from the burrow at about three weeks old.

Brown rats are common in sewers. Burrowing rats may cause the collapse of an old sewer, where the cracked brickwork offers many nesting holes.

Brown rats are suspicious of any unfamiliar objects, and approach warily, sniffing all the time. They may take days to eat new bait or venture into a trap.

The young are born naked and blind in a nest of loose straw, shavings, rags or similar material. A litter usually numbers about seven, but a big female may have up to 11.

Small, finely
haired ears

Thick,
scaly tail

Brown rats like to live where there is undisturbed
shelter and water close by. They are usually active at
night, but may sometimes be seen during the day.

Brown rat *Rattus norvegicus*

A widespread pest, the brown rat fouls food stores and gnaws stored goods. It spread to Britain in the early 18th century, probably from Russia, and is found not only in buildings but also in many places out of doors, such as rubbish tips and muddy shores where debris is washed up by the tide. Very large populations often build up in farms, where there is plenty of food even though combine harvesters have done away with the corn ricks that once harboured thousands of rats. In a fine summer, young rats may spread into hedgerows and lay-bys far from buildings. Brown rats are also at home in sewers, where they are especially liable to pick up and transmit diseases.

Where food and shelter are abundant, a female brown rat can produce five litters a year totalling 50 young, all able to breed at three months old. But many young rats are caught by cats, owls, foxes and other predators. Rat numbers are also controlled by the use of special poisons that prevent blood from clotting – although some rats have developed a resistance to them in recent years. The white rats of pet shops and laboratories are a specially bred form of brown rat. Unlike their aggressive wild cousins, they are tame and inoffensive.

Common around buildings,
in farms and hedgerows. Less
usual on high ground, moors.

The brown (or common) rat will eat
anything and thrives where there are
food stores or waste. It has coarse,
grey-brown fur, small finely haired ears
and a thick, scaly tail always shorter
than its head and body. Male up to
11 in. (28 cm) head and body, 9 in.
(23 cm) tail. Female smaller.

133

Not all black rats are black. Brown forms are common, but can be distinguished from brown rats by their large pink ears and long tails. Brown varieties may have a grey belly or a creamy-white belly.

Grain and food stores used to be raised on pillars of mushroom shape so that rats could not climb in.

Like all rodents, the black rat wears down its constantly growing front teeth by gnawing, and may cause expensive damage to woodwork. It will even gnaw lead pipes and electric cables.

Its large eyes and ears and long whiskers help the black rat to find its way in dark buildings. It eats all sorts of food, but particularly cereals.

Like the brown rat, the black rat has greasy fur that discolours a surface it rubs against regularly. A rat hole gnawed through wood can be recognised by the stained edge.

Black rats are active at night. They climb easily – up a rope, brickwork or even a thin strand of wire. But, unlike brown rats, they are reluctant to enter the water.

Large
pink ears

The black (or ship) rat is most likely to
be seen in a dockside warehouse. Its
large, bare, pink ears are striking, and
its thin tail is always as long as, or
longer than, its head and body. Male
up to 8 in. (20 cm) head and body, 9 in.
(23 cm) tail. Female smaller.

Fast and very agile, the black rat runs easily down a
smooth, vertical surface such as a window frame. Its
tail is lightly curled to assist balance.

Almost extinct. Found mainly
inside buildings, chiefly in
major ports and old towns.

Long
thin tail

Ship mooring lines
used to be fitted with
cones to stop rat traffic
between a ship and
the shore.

Black rat *Rattus rattus*

Widely blamed for carrying plague, the black rat thrives in a
man-made environment, and was numerous in the untidy and
insanitary towns of medieval Europe. The rat originated in
tropical Asia, but spread to Europe as trade between countries
developed; it reached Britain probably in the 11th century.
Plague was rife in Asia, the virus being transmitted by the rat flea.
During the 14th century, a massive outbreak of disease killed
some 25 million people in Europe, including more than a third
of Britain's population, and was thought to be plague, spread by
the rat flea. But modern research suggests that the disease may
not have been rat-borne plague after all.

Because of its tropical origins, the black rat in Britain prefers
to live in warm, sheltered buildings. Since the coming of the
larger, more aggressive and adaptable brown rat some 300 years
ago, the black rat has disappeared from most of its old haunts. It
is now probably Britain's rarest mammal. Few have been reported
in recent years, and the black rat may soon be extinct in Britain.
Black rats usually have from three to five litters of six or seven
young a year; they mature at about four months old. Humans and
domestic cats are their chief enemies.

135

A tennis ball fixed about 18 in. (45 cm) off the ground makes a nest-box suitable for a harvest mouse.

There are usually from three to eight young in a litter, born on a layer of chewed grass leaves. Females are pregnant for 17–19 days and may have three litters in a year, each in a new nest.

Seeds, grain, grass shoots and soft fruit, as well as insects such as weevils, are all part of the harvest mouse's diet. Because it is very active, it has a large appetite for its size.

In winter the mouse's coat is a darker brown. The winter nest is built low, usually in a clump of grass or under a hedge.

Breeding nests are built well off the ground, often in grass or reeds, and are spherical – about 2½ in. (64 mm) across. The framework is woven from neatly shredded living leaf blades, so the nest hangs between stems. Non-breeding nests are more loosely built.

When tall plant stems die back in autumn, the harvest mice are left exposed and seek cover in low vegetation or sometimes in barns. Many mice die during winter.

Its blunt nose, small ears and yellow-brown fur distinguish the tiny, agile harvest mouse, which lives among tall, stiff-stemmed vegetation and uses its tail as a fifth limb to grasp stalks. About 2½ in. (64 mm) head and body, 2½ in. (64 mm) tail.

Blunt nose

Small ears

Tail grasping stalk

Fairly widespread in coarse vegetation and hedgerows. Not found on high ground.

Harvest mice breed from late May to October. A male courting a female runs after her making a chattering noise. Sometimes she turns round and fights him off.

Harvest mouse *Micromys minutus*

One of the world's smallest rodents, the harvest mouse weighs less than a 2p coin. Because it is so tiny it can climb fast and confidently among thin stalks. As their name implies, harvest mice have always been associated with cornfields, but now that these are reaped by machines and then ploughed up, it is difficult for them to live there permanently. They more often live in the long grass at the base of a hedge, invading the growing corn crop to nest and feed in early summer. Tall vegetation such as brambles and rushes is, however, their main summer home, and they are common in reed beds despite the water below. The peak breeding season is August and September. When young harvest mice become independent at about 16 days old they are grey-brown, and later moult to adult colouring.

Although harvest mice are threatened by urban development and farming methods such as stubble burning and scrub and hedge removal, they may be able to benefit from the huge, undisturbed areas of rank grass that are developing along motorway embankments. Because they are active by day and night, the mice are likely to be caught by both daytime and night-time predators, and most live for only a few months.

137

Where there is plenty of food, house mice may live in fields and hedgerows in summer. Some may also move outdoors in late summer when indoor populations increase to high numbers. But few survive cold weather outdoors, and most spend the winter in a building.

A mouse often sniffs with its nose raised. Its sense of smell is acute, and it can find its way about and recognise other members of its family by scent.

Extremely adaptable, house mice can live in all sorts of places. Some even thrive in frozen meat stores, and have developed longer fur to keep out the cold.

A bar of soap can provide a meal for a house mouse. It will eat all sorts of unusual items, although cereals and fats are its favourite food. It rarely takes liquids.

Male mice fight to establish social dominance or defend their territory, and any adult in a group may attack an outsider. Fighting occurs more often as the population increases.

Pointed nose

Greasy fur

Scaly tail

Many kinds of stored food will attract the house mouse, which is mostly active at night. Alert and nimble, it has prominent ears, big eyes and a pointed face. Its grey-brown coat is greasy and glistening and its tail long and scaly. Unlike other mice, it is smelly. About 3¼ in. (83 mm) head and body, 3¼ in. (83 mm) tail.

Anything soft and easily shredded, such as grass, sacking or newspaper, may be used to make a nest. One female can produce about 50 babies a year.

House mouse *Mus musculus*

Unlike other mice, the house mouse has a strong smell and greasy fur. It taints the places it lives in – homes, warehouses, hospitals and other public buildings. Mice also leave their black droppings and their urine about, carry diseases and parasites, and cause damage by their gnawing. Although house mice live for only about 18 months, and many die in their first winter, it is hard for the humans they live with to get rid of them. They breed so fast that in one summer the mice from one nest will have multiplied many times over. Females bear from five to ten litters a year with five or six young in a litter. The young leave the nest at three weeks old and three weeks later the females among them are ready to breed.

House mice live close to their food supply and move on only when food is short. They are easily transported accidentally among food or goods and, with the spread of humans, have colonised most of the earth from their first home in Asia. The species essentially lives with man, indoors. When the Scottish islands of St Kilda were deserted in 1930, the house mouse – probably a resident since Viking times – became extinct within a few years.

Widespread, mostly in and around buildings, but also in hedgerows and on farmland.

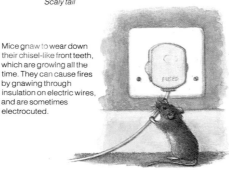

Mice gnaw to wear down their chisel-like front teeth, which are growing all the time. They can cause fires by gnawing through insulation on electric wires, and are sometimes electrocuted.

139

Always cautious, the mice sniff suspiciously before approaching anything unfamiliar. They have an acute sense of smell and rely chiefly on odours as a means of recognising other mice.

Autumn fruits such as hawthorn and other berries are part of the wood mouse's diet. It climbs well, and often uses places such as an old bird's nest high on a tree branch as a feeding place.

The wood mouse is at home in all sorts of places besides woodland, including sand-dunes. A male regularly forages over a territory roughly half the size of a football pitch.

Seeds are the wood mouse's main food, along with shoots and buds. It also eats snails, nibbling its way through the shell, and a variety of insects and their larvae.

Wood mice have large hind feet that enable them to leap away like kangaroos. They usually move very fast, often bounding with their front feet tucked up.

The mice dig their own burrow systems, where they store food and spend the day. Young mice are born there in a nest chamber.

140

Large ears

Yellow streak

Long tail

Its sandy-brown coat and large ears and eyes help to distinguish the wood mouse. The underparts are white, with a yellow streak on the chest, and the tail is lightly haired. About 3¾ in. (95 mm) head and body, tail longer.

The wood mouse is common but not often seen, as it is active only at night. It prefers dark nights, and uses its large eyes and ears for finding its way about.

Wood mouse *Apodemus sylvaticus*

Probably the most widespread and abundant British mammal, the wood mouse – also known as the long-tailed field mouse – is not confined to woodlands. It thrives equally well in more open places, even on moorlands and mountain-sides, and is also common in gardens, where it often lives near sheds and out-buildings and causes people to think house mice have moved in. However, unlike house mice, wood mice do not smell strongly.

The wood mouse is very active, running and bounding from place to place and venturing into open places where other small mammals will rarely go. Although it moves only under cover of darkness, it is frequently taken by predators, especially owls and cats. Most wood mice stay in the same general area but may travel a quarter of a mile (400 m) in one night. In winter they sometimes go into a torpid state – almost like hibernation – in which they use far less energy than usual. This helps them survive food shortages. The population is at its lowest at the end of winter, but numbers soon build up. Breeding starts in March, and a female may bear four litters, each of about five young, in a year. Wood mice are generally short-lived; the maximum life-span in the wild is rarely more than two years.

Widespread in all habitats from woodland to sand-dunes, mountain-sides and gardens.

When mice are frightened, they often wash and groom themselves. They sit on their haunches and lick vigorously at their armpits, forelimbs and belly.

141

Small woodland mammals are preyed on mainly by tawny owls. When mice and vole numbers are low, an owl may have difficulty in finding enough food to rear even one chick. Below its roost lie regurgitated pellets – wads of indigestible fur and bones from its prey.

Bank voles forage among thick cover by day or night. They eat berries and seeds, nibble at fungi and take some insects.

Mice, voles and squirrels nibble the caps of fungi such as *Boletus* at the edge, and leave fragments of the flesh lying about. At night slugs may also feed on the cap, but make neat, rounded holes in it.

The common dormouse ventures out only at night to forage mainly among tree branches. In autumn it fattens up before hibernating about October in a nest built at or below ground level.

The woodland floor at night

Many kinds of mammal, large and small, make their homes in deciduous woodland. Not only do the trees and shrubs provide more shelter from winds and cold than open country, but for mammals that can climb they give extra living and foraging space among their branches. Above all, the different kinds of trees, shrubs and flowers offer a wide assortment of food for the mammals and also for a vast number of insects, which are in turn another source of food for the mammals. A wood covering about $\frac{1}{4}$ sq. mile (65 hectares) can support more than 5,000 mice and voles as well as many shrews, moles, squirrels and maybe some badgers and deer.

Predators such as tawny owls and weasels are attracted by the abundance of small mammals. The night-flying tawny owl, especially, depends for food on the mice and voles that forage on the woodland floor. In summer when undergrowth is dense and small mammals well hidden it often hunts over open fields and hedgerows. But when the vegetation dies back in late autumn and winter, woodland mammals are easier prey. The owl claims a woodland territory ready for breeding time about March; territory size is largest when mice and vole numbers are low.

Wood mice feed on more open parts of the woodland floor than bank voles. Several mice will share a space, showing little aggression over the nuts, fruits, seedlings and insects it offers, except during the breeding season.

143

The nest of grass and leaves is made underground within the burrow. There are five or more babies in a litter. They emerge from the nest when they are about 18 days old.

Like wood mice, yellow-necked mice eat mainly seeds and fruit, such as acorns, hazel nuts and blackberries. In summer they also feed on insects and small animals such as snails and spiders.

Good climbers, yellow-necked mice will search at the top of a tree for food such as new buds. They have been found as high up as 33 ft (10 m).

In autumn, yellow-necked mice often enter outbuildings or houses in search of food and shelter. They are sometimes caught by cats or in traps.

Better climbers than wood mice, yellow-necked mice are more likely to be found in an attic. Stored apples provide the mice with a good winter food supply.

Like the wood mouse, the yellow-necked mouse comes out only at night. In autumn its food supply includes hawthorn berries. It may live about two years.

Yellow collar

Orange-brown flank

Found in woods, hedgerows, but distribution is patchy. Much rarer than wood mouse.

The yellow-necked mouse looks very much like the wood mouse, but is distinguished by its distinct yellow collar. It is also bigger and heavier than the wood mouse, and its sandy-brown coat has more orange on the flanks. About 4 in. (10 cm) head and body, tail usually longer.

Yellow-necked mouse *Apodemus flavicollis*

Not only does the yellow-necked mouse resemble a large, sandy-coloured wood mouse, but it is very like the wood mouse in its behaviour. Both mice are strictly nocturnal, with the big ears and eyes of creatures that need to pick up in the dark the faint sound or slight movement that warns of approaching danger, and both climb well and often search for food among high branches. Both are also found in woods, hedgerows and gardens. This seems to contradict a basic biological principle that no two animals can live in the same place and share the same food supply without one ultimately displacing the other. The yellow-necked mouse might be expected to become dominant as it is larger, but it is the wood mouse that occurs over most of Britain. Yellow-necked mice occur only in the south, and even there huge tracts of suitable woodland are apparently without them. Where they do occur, yellow-necked mice may increase to considerable numbers and then inexplicably disappear a year or two later.

In parts of the south-east, where yellow-necked mice can be quite common, they often go into gardens and even houses in autumn – perhaps seeking a dry, sheltered place for the winter.

145

As it swims, the vole's blunt nose is held clear of the water at the tip of a V-shaped wave. Voles may fall prey to herons, owls, pike or larger mammals such as mink.

The voles live in a system of burrows in a waterside bank. The entrances may be above or below water. They swim by paddling with all four legs and are kept warm and dry by a short dense undercoat below their long outer fur.

Active by day, the vole moves among grasses, sedges, willow shoots and other waterside plants. It grasps them in its forepaws to eat the best parts, leaving a trail of fragments and stumps.

A nest of woven plant stems is sometimes made at the base of sedges on marshy ground. Usually the nest is below ground in a burrow system. Four or five litters of about five young can be reared between March and October.

Chubby face

Long, glossy fur

Although the water vole is about the same size as a brown rat, it differs in having a chubby face with a blunt nose, short furry ears almost hidden by its long, glossy, dark chocolate-brown fur, and a shorter tail. Male 8 in. (20 cm) head and body, 4¾ in. (12 cm) tail. Female slightly smaller.

Becoming rare in Britain and Ireland. Found by ponds and slow-running rivers.

Completely black water voles occur, especially in Scotland. Voles with a white tail-tip are also more frequent in Scotland.

The distinct plop it makes when diving into the water often reveals a water vole's presence. In the water it resembles a brown rat, but has a blunter head.

Water vole *Arvicola terrestris*

Water voles are often called water rats, but are only distantly related to rats. Unlike brown rats, water voles prefer clean water in relatively undisturbed areas by lowland river banks or the fringes of ponds and lakes. Their numbers are declining rapidly in many areas, partly as a result of predation by an expanding population of mink, and they are now very rare.

The harmless water vole feeds almost entirely on waterside plants, and spends most of its life in a narrow strip of land at the water's edge. Small heaps of droppings mark the limits of its home range. A male occupies about 140 yds (130 m) of bank, and often stays in the same area a long time, sometimes all its life. A female occupies only half the range of a male and will sometimes leave her regular haunts to live elsewhere. Young voles, which are very dark brown with a long, almost black, tail, are sometimes found away from water in damp woodland and grassy areas; they may be mistaken for field voles, although bigger and darker. Movement from a population often occurs when numbers are high after a good breeding season, or when shallow ponds are drying up because of dry weather. A water vole's normal life-span is 12-18 months.

Hazelnuts are frequently eaten. The sharp-toothed bank vole gnaws a hole in the shell and takes out the kernel in small pieces.

Seeds, berries, nuts, fruit, green plants and fungi are all part of the bank vole's diet. Food may be stored underground or taken there to eat in safety.

Where there is sufficient thick undergrowth, the bank vole forages busily by day or night along a network of tunnels beaten through vegetation or dug underground.

Skomer vole

The Skomer vole found on Skomer Island, Dyfed, is twice as heavy as mainland voles and about 4¼ in. (11·5 cm) long in head and body. It is one of four larger island sub-species; the others are on Jersey, Mull and Raasay.

There may be four or five litters, each with four or five babies, between April and September. The nest is sometimes above ground, perhaps in a tree crevice, but often up to 4 in. (10 cm) below ground in a chamber reached by tunnels.

Small ears

Blunt nose

Only about 1 oz (28 g) in weight, a bank vole can climb delicately among bramble stems and balance on a sideshoot as it reaches out for a berry.

The bank vole can be distinguished from a mouse by its chubby appearance, blunt nose, small eyes and ears and short, furry tail. Adults have a glossy, chestnut-brown coat that may shade to grey on the belly. About 3½ in. (90 mm) head and body, up to 2⅜ in. (60 mm) tail.

Short tail

Commonest in lowlands. Found in woods, hedges and scrub. Very limited in Ireland.

The bank vole has a redder coat and more prominent ears than a field vole. Young bank voles are grey-brown and more difficult to tell from field voles.

Bank vole *Clethrionomys glareolus*

After the wood mouse the bank vole is probably the most abundant of Britain's small rodents. It is more likely to be seen during daylight than the wood mouse, and tends to run and scurry rather than move in leaps and bounds. Although it may sometimes be found in long grass, wet places or on mountainsides, the bank vole much prefers to live where there is dense cover. It is rarely found far from bramble thickets, hedgerows and other woody scrub, and is common in country gardens. Each bank vole occupies a home range, and does not normally venture more than about 55 yds (50 m) from its nest. Males generally range more widely than females.

In mild years when there is plenty of food available, bank voles may begin breeding early and continue well into late autumn. A vole born early in the year may itself be raising a family within a few weeks, so the population builds up quickly. But fewer than half of those born survive the first few months. After they leave the nest at about 18 days old, young voles are often taken by predators such as weasels, or may die during cold, wet weather. The more robust survivors may live for 18 months. Bank voles have been found in Ireland only since 1964.

Field voles often nest under logs and other objects lying in dry grass. If the covering is lifted, a nest chamber and runway system are revealed among the grass stems. Voles will also nest under a blanket of snow.

Field voles are very plentiful where there is lush, undisturbed grass. Ideal places are young forestry plantations.

Orkney vole
Microtus arvalis

Found only on the Orkney Islands in Scotland and the Channel Island of Guernsey, the deep brown Orkney vole is very like the field vole in habits and looks, but is heavier. It weighs about half an ounce (10–15 g) more. About 4½ in. (11·5 cm) head and body, 1¾ in. (45 mm) tail.

Four or five litters of from four to six young are reared between March and December. By ten days old they are furred and by 16 days old they are weaned. At six weeks old, young females are ready to mate.

Grass is the field vole's main food, particularly the succulent lower stems. It will also eat bulbs, roots and tree bark at ground level.

Yellowy-brown fur

Field voles are the main food of barn owls, forming 90 per cent of their diet. Other animals that take the voles include kestrels, foxes, stoats, weasels and snakes.

Short tail

Its yellowy-brown colouring helps to distinguish the field vole from the red-brown bank vole. It is chubby like the bank vole, with a blunt nose and short ears, but it has a shorter, pinker tail and is also known as the short-tailed vole. About 4 in. (10 cm) head and body, 1½ in. (40 mm) tail.

Field vole *Microtus agrestis*

Overgrown fields and places with long, rough grass are typically the home of the aptly named field vole, which particularly likes damp, tussocky grass. Aggressive and noisy, field voles utter loud squeaks and angry chattering noises as they defend their small territories, driving out other voles. Each vole makes runways among the grass stems, usually centred on the tussock where it nests; it feeds frequently, by day or night.

Field voles are taken by a host of predators, but are prolific breeders. Populations in a favourable habitat often increase rapidly to number in their thousands – a vole plague. The plague is followed by a rapid decline, probably due to less successful breeding because of overcrowding and heightened aggression. These high and low populations occur at intervals of from three to five years, often accompanied by similar fluctuations in predator populations. Field voles are very abundant for a few years in young forestry plantations, but as the trees grow they cast a dense shade and the grass dies, forcing the voles to go elsewhere. Some survive on the grassy fringes, from where they can quickly recolonise grassy areas that develop once trees are felled. The normal life-span is about a year.

Widespread in grassland and hedgerows in both lowlands and uplands. None in Ireland.

Adults are very belligerent. They compete for territory and are quick to fight in its defence.

151

Most of the squirrel's time is spent in the treetops. It is found chiefly among conifers, usually in extensive forests. On some trees squirrels strip bark to reach the sap.

Although it does not come out in wet weather, the squirrel does not hibernate and is often seen in midwinter. It cannot go more than a few days without eating.

In autumn when food is plentiful, squirrels forage by oak or beech trees for acorns, beechmast and other foods to store for winter. They also eat heartily in autumn to put on fat reserves for winter.

The red squirrel uses its strong, sharp claws to grip the bark as it runs head first down a tree. It is strong enough to hang by just one foot. Its claws often scar the tree bark.

True albinos, with pink eyes, are sometimes seen in Scotland. Very dark brown varieties also occur.

Bushy
tail

Ear-tufts

Chestnut
fur

Early on winter mornings, red squirrels are likely to be
seen high in the trees searching for food such as larch
cones, from which they take the seeds.

The squirrel's ear-
tufts are especially
long in the winter
(particularly on an
adult), and give it a
perky appearance.

Declining; mostly gone from
south. Mainly restricted to
large areas of mature forest.

The long, silky winter coat is
chocolate brown, the fur on the
back sometimes looking dark grey.
The tail is a uniform dark brown.

In summer the attractive red
squirrel has bright chestnut fur
with orange-brown feet and
lower legs. Its ears are tufted,
but tufts may be smaller or
absent in young animals. As
summer progresses, the
squirrel's bushy tail bleaches
to a pale cream. About 8 in.
(20 cm) head and body;
7 in. (18 cm) tail.

Red squirrel *Sciurus vulgaris*

Red squirrels are most likely to be seen in heavily forested areas
soon after dawn. Mature Scots pine woods are a favourite
habitat, but they are also found among other conifers such as
larch and spruce. These trees provide a high thoroughfare
among the branches and year-round food from cone seeds,
buds, shoots and pollen, although the squirrels also like to
forage for nuts among nearby deciduous trees.

The Scots pine was the only large native conifer to survive
the last Ice Age, and the red squirrel one of the last mammals to
colonise Britain before it was cut off from the rest of Europe
some 9,000 years ago. Until the 1940s the animal was fairly
widespread. Now it has disappeared from large areas of the
country and its place has been taken by the grey squirrel. The
reasons for the red squirrel's decline are not clear. One cause
may be the extensive loss of suitable woodland, and the
clear-felling of large areas of conifers. It is unlikely that grey
squirrels drive away red squirrels, but they may stop them
repopulating wooded areas by competing for food. They live for
about three years. Few predators can catch them in the trees; on
the ground they may be killed by foxes, birds of prey or cars.

153

Squirrels recognise food by its smell. They enjoy a wide variety of fruits, nuts, pollen, flowers and fungi, and thrive best where a mixture of different tree types gives the fullest range of food.

The mating season often begins on warm days in January. Courtship is long-drawn-out and includes chases through the branches. A female may have several mates in one season.

Tree seeds and red squirrel survival

Red squirrel numbers from season to season depend on the seed crop of the dominant trees where they live. Where there are plenty of pine cones, hazel nuts or similar food, red squirrels build up generous fat reserves and many survive the winter in good condition. Unlike the grey squirrel, however, they cannot digest acorns. They begin breeding early in the following year, and many babies are born and reared. In years when tree seeds are less abundant, the squirrels begin winter with poor fat reserves and many die from starvation or disease. Most of the survivors are not fit enough to breed successfully. The population can take years to recover, and while red squirrels are few, grey squirrels often move into the area. The struggling red squirrels are then forced to compete for food and space with a larger animal.

A litter of red squirrels usually numbers three or four, sometimes up to six, but in larger litters more die. Older females may produce two litters by the end of the breeding season in August. Males take no part in rearing young. A red squirrel is most in danger from predators for the first few months of life, after leaving the nest and before losing its fluffy, juvenile coat and becoming fully independent.

Food is held in the forepaws. Squirrels bite the scales off ripe pine cones to get at the seeds. Chewed cone cores and discarded scales litter the ground under the trees.

If the breeding nest is disturbed, a mother will carry her babies in her mouth one by one to an alternative nest near by.

Youngsters leave the nest when about two months old. Soon afterwards they become independent of their mother and develop an adult coat.

Young squirrels are born in a specially built drey with a thick, grassy lining inside a framework of twigs. It is about 12 in. (30 cm) across and usually wedged against a tree trunk. Young are born blind and naked, but are fully furred by about three weeks old and open their eyes at four weeks old.

A loud churring noise in the tree-tops often reveals the presence of grey squirrels as they aggressively scold or chase off an intruder.

Agile grey squirrels can run along slender twigs and leap from tree to tree. Young ones soon learn to do so. If they fall they can land safely from heights of about 30 ft (9 m).

White squirrels are fairly common, in the south-east especially. They have pink eyes and are true albinos.

Squirrels are active by day. They like to sit upright on vantage points to survey their surroundings, relying as much on eyesight as smell for information.

Silky black squirrels are sometimes seen, although rare. They are found mainly in and around Bedfordshire, where black varieties from North America were released early this century.

Winter fur is dense and a bright, silvery grey with a brown tinge along the middle of the back. It is replaced by brownish summer fur during April and May.

Bushy, grizzled tail

No ear tufts

The grey squirrel does not hibernate, and may be seen feeding on fine days during the winter. A half coconut fixed to a tree provides it with a feast.

In summer the grey squirrel's fur is often yellowish-brown, on the flanks and feet especially. It has a bushy, grizzled tail that stays the same colour all year, white underparts, and ears without tufts. About 10 in. (25 cm) head and body, 8 in. (20 cm) tail.

Yellowish-brown fur

Common in forests, gardens, hedgerows: scarce on high ground. Range expanding.

Silvery winter fur begins to grow in autumn, starting on the rump. The complete moult takes up to six weeks.

Grey squirrel *Sciurus carolinensis*

One of Britain's most familiar and frequently seen mammals, the grey squirrel is a native of the hardwood forests of the eastern United States. It was introduced to this country in the mid-19th century, but did not become established in the wild until about the turn of the century, after many releases.

Unlike native red squirrels, grey squirrels can live happily in hedgerow trees, parkland, gardens and other places without large areas of trees. Britain's patchwork countryside with dotted trees and isolated copses suited them well and they became abundant. Where grey squirrels spread to live alongside red squirrels, the greys usually became more numerous after a while and displaced their smaller cousins, generally without obvious signs of conflict.

As well as nuts, acorns, beech-mast and fungi, grey squirrels eat tree bark, leaves, shoots, buds and flowers. In commercial woodlands they damage valuable trees and are a serious pest. Elsewhere most people find their behaviour attractive and endearing. Because the squirrels have few predators, some live for up to 10 years. But most die much younger, from starvation, accidents such as forest fires or as a result of pest control.

157

Instead of building itself a drey, a grey squirrel may make its home in a hollow tree or a woodpecker's old nest-hole.

A summer drey may be flimsy and lodged amid small branches.

Usually a winter drey is on a sturdy branch close to the trunk.

The football-sized nest, or drey, is made of twigs, often with the leaves still attached. It is built fairly high in a tree and lined with dry grass and shredded bark.

The grey squirrel at home

Because grey squirrels are agile, acrobatic and active by day, they are one of the most interesting and easiest to watch of all British wild creatures. All sorts of behaviour can be observed – they can be seen sitting surveying their surroundings, their twitching tails indicative of uneasiness or suspicion, or they can be aggressively grinding their teeth and chattering loudly at an intruder or predator. They can also be seen as they run swiftly up and down trees, climb walls or bound briskly across the ground in short leaps.

Intelligent and resourceful, grey squirrels have been particularly successful in invading towns and gardens where there are trees. They quickly become quite bold, and soon learn to raid bird tables for extra food, or to chew through the string holding a nut basket suspended and collect their booty when it falls to the ground. Not only do they benefit from food put out for birds, they will also occasionally raid nests to steal eggs and nestlings. Young squirrels are born in tree-top dreys in spring or early summer. The average litter size is three, and the male plays no part in the rearing. The youngsters disperse when their teeth are fully grown and they can feed themselves, usually at about ten weeks old. They breed for the first time at a year old.

The squirrel uses its forepaws to manipulate food, especially tough things such as nuts, which need a lot of gnawing to open. Nutshells litter the ground at a favourite feeding site.

Strong feet and sharp claws enable grey squirrels to climb brick walls to reach bird food left on upstairs window sills. They will also climb to attics or the roofs of outbuildings to nest.

Squirrels soon master the art of reaching a basket of nuts put out for birds, even if it is suspended by wire from a clothes line.

Tree bark is gnawed to get at the nutritious sapwood below. The raw scar left on the tree may lead to distorted growth and encourage fungal attack.

In late winter, courting squirrels often frisk and chase across lawns or along branches and logs. They run fast, with their tails flicking and their legs outstretched.

Surplus food is stored for winter by burying. Often the cached food is not recovered, and acorns and other seeds sprout the following year, so aiding tree dispersal.

159

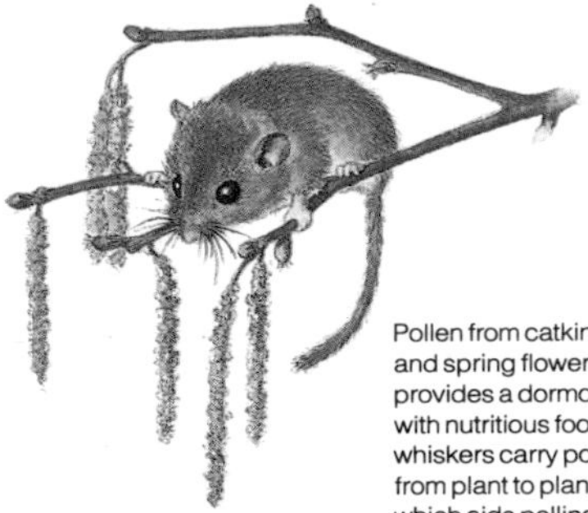

Dormice spend more time than other mice climbing in bushes and trees, where they obtain all their food, including many insects.

Pollen from catkins and spring flowers provides a dormouse with nutritious food. Its whiskers carry pollen from plant to plant, which aids pollination.

Hazel nuts are often eaten. The dormouse makes a hole in the side of the shell to get at the kernel. The cut edge of the hole has toothmarks along it.

A domed nest about 6 in. (15 cm) across is built in a bush, often a thorny one to deter predators. The nest may be sited several feet above the ground.

Between May and September there is one litter, sometimes two, of from two to seven young, born blind and naked. They soon grow grey fur, but moult this before leaving the nest at a month old, when they resemble adults in colour, but are greyer.

To build a nest, the dormouse strips bark off shrubs such as honeysuckle. It also uses dry leaves and grass.

Orange-yellow fur

Secretive, nocturnal and rarely seen, the common dormouse is distinguished from all other mice by its fluffy tail, orange-yellow fur and chubby build. It also has smaller ears than other mice, but much bigger eyes than voles. Its belly fur is creamy-white. About 3 in. (75 mm) head and body, 2½ in. (64 mm) tail.

Fluffy tail

A winter nest is built on or near the ground, among tree roots or in a hedge bottom or coppice stool. Here the dormouse hibernates alone from October to April.

Uncommon. Found in southern hedges and woods. May occur in parts of northern England.

An agile climber, the dormouse will live in a bird's nest-box, even one fixed 12 ft (3·6 m) above the ground.

Common dormouse *Muscardinus avellanarius*

In times gone by, the common dormouse was a familiar animal to country folk, and frequently kept as a pet. It was also called the hazel dormouse because it was often found in hazel coppices. Yet today common dormice are rarely encountered, and no one is sure whether this is because they have become rare and threatened, or are merely overlooked. In the days when men spent many hours hedging and ditching – trimming hedges and cleaning out ditches – many a drowsy, hibernating dormouse was exposed, for these tasks were carried out mainly in winter. Many dormouse nests were also discovered when the poles were harvested from coppiced trees. Today these tasks, if done at all, are usually performed by machines, so the presence of dormice is much less likely to be noticed.

In summer dormice live in woodland shrubs and bushes or among tall hedges or dense scrub. They come out only at night to feed on insects, nuts and flowers, so are rarely seen. As they stay hidden among the branches they are not often caught by cats, nor do they normally venture into traps. Dormice may be taken by crows, magpies, owls and foxes, but can live to be about five years old in the wild. Some pets have lived to be six.

Its tail helps the dormouse to balance. It is whisked about as the agile animal climbs among small branches in the tree-tops.

From October to April the fat dormouse hibernates. It stays in its nest all the time and lives off its fat reserves, losing nearly half its weight.

Fat dormice gnaw trees, particularly those with sweet, juicy sapwood under the bark. Fruit and timber trees may be severely damaged.

A single litter of four or five young is produced each midsummer in a nest made of dry leaves and grass in a tree hollow or an old squirrel drey, or sometimes in the corner of a loft.

Sometimes the dormouse eats birds' eggs and nestlings, large insects and any small animals it can catch.

Plant material is the main food, particularly fruits but also nuts, bark and fungi. The dormouse holds food in its forepaws while eating.

Dark eye rings

Grey body fur

In late summer the fat dormouse eats heartily to prepare for hibernation. By autumn it may have nearly doubled its summer weight of about 5 oz (140 g).

Bushy grey-brown tail

The bushy-tailed fat dormouse looks like a small grey squirrel. Its belly is white, and its fine grey body fur often has darker, brownish tinges on the tail, the outside of the legs and in a ring round each eye. About 6 in. (15 cm) head and body, 5 in. (12·5 cm) tail.

Mainly confined to woodland in the Chilterns. It often winters in houses and sheds.

Fat dormouse

Grey squirrel

Although the fat dormouse may sit up like a squirrel, it keeps its tail laid flat. Fat dormice come out only at night, and grey squirrels are active only during the day.

Fat dormouse *Glis glis*

The Romans used to keep fat dormice in captivity and deliberately overfeed them to make a succulent and unusual meal, so they are also known as edible dormice. In the wild these squirrel-like dormice eat extra food in late summer to fatten themselves and build up reserves for the winter.

The fat dormouse was not introduced to Britain until recent times. A few were brought from the Continent, where they are common, and released at Tring in Hertfordshire in 1902. Now they are found in many woodland and suburban areas of the Chilterns, and although in some places they are numerous enough to be regarded as pests, they have not spread very much. The fat dormouse is not a great wanderer and rarely seems to travel more than a few hundred yards from its nest. It spends most of its time high in tree branches, foraging only at night, so is not often seen. But signs of its activity are more evident; it damages trees by chewing bark, buds and growing shoots, and in autumn may enter a house or shed and gnaw woodwork or stored food. It also comes indoors to a dry place to hibernate, and may keep people awake by scuttling about in the loft. It makes loud wheezing and churning noises, calling from the trees.

163

Indian giant squirrel
Ratufa indica

In the winter of 1959–60, an escaped giant squirrel lived wild in Blackheath, London, for several weeks before capture. No one knew how it got there. About 11 in. (28 cm) head and body, tail same length.

Crested porcupine
Hystrix cristata

During the 1970s, a pair of crested porcupines escaped from a zoo in Staffordshire and lived free for about two years. At the same time there were some escaped Himalayan porcupines living free in Devon. All damaged trees, but did not seem to multiply. About 25 in. (64 cm) head and body.

Mongolian gerbil
Meriones unguiculatus

The hardy, soft-furred gerbil from the desert regions of Mongolia has been a popular pet in Britain since the 1960s. Many have escaped and established wild colonies, often under sheds and outbuildings. About 4 in. (10 cm) head and body, tail same length

Alien rodents at large

Pets and zoo animals occasionally escape and may be seen in the wild. Even large animals such as zebras and sea lions sometimes get away, but are usually quickly recaptured. Small, agile rodents such as squirrels easily elude capture and may live wild for years. Occasionally small animals escape in a group and establish a breeding colony, as hamsters and gerbils have done many times, adding exotic species to Britain's wildlife. Sometimes people deliberately release pet animals into the wild; this is cruel to animals that are unused to fending for themselves.

Most alien animals do not survive very long in the unsuitable conditions of a strange country, but some become well established and increase rapidly. They may then become pests that are hard to get rid of. Escaped coypus and musk rats have shown the damage that can be done by alien species that become established in our countryside. To deliberately release animals from abroad into the wild is now illegal.

Musk rat
Ondatra zibethicus

The semi-aquatic musk rat was brought to Britain from North America in the 1920s for fur farming. Some animals escaped to form extensive colonies, and did a lot of damage to crops. By 1937 the animal had been exterminated. About 12 in. (30 cm) head and body, 10 in. (25 cm) tail.

Golden hamster
Mesocricetus auratus

A stoutly built burrower from eastern Europe and the Middle East, the soft-furred, nocturnal hamster feeds on fruit, vegetables and grain. It stores food in its large cheek pouches for eating at leisure. About 7 in. (18 cm) long including tail.

West Indian mouse
Nyctomys sumichastri

Dark-eyed, furry-tailed mice have reached Britain from the West Indies with cargoes of bananas, which are part of their diet. They normally arrive singly, and have not established colonies. About 4½ in. (12 cm) head and body, tail same length.

Unless kept in a secure metal cage, hamsters will gnaw and burrow their way to freedom. They have been known to burrow out of a shop basement and tunnel under the pavement.

In March and April particularly, hares indulge in energetic leaping and wild chases – hence the saying 'mad as a March hare'. They breed at any time of the year.

A leveret is born in the open with a full coat of fur and its eyes open. It lies low in a form – a depression made amid long grass. A litter usually numbers two or three, each kept in a separate form.

During the day the hare rests crouching low against the ground in scrub or grass or in a ploughed furrow. Its ears are laid flat and from a distance it looks like a large clod of earth or turf.

If disturbed the hare can run as fast as 35 miles (56 km) an hour – it runs with its tail held downwards, showing the dark topside. Normally it lopes along unhurriedly, its long hind legs placed in front of its forelegs at each stride.

Hares eat grass and other plants, but farm weedkillers destroy many of the plants they need. They also eat their own soft droppings.

Black-tipped
ears

A shy, alert animal of open country, the
brown hare has black-tipped ears
about 4 in. (10 cm) long, large staring
eyes and powerful hind legs. The fur on
throat and flanks is orange. Male (Jack)
22 in. (55 cm) long. Female (Jill) only
slightly smaller.

Staring
eyes

Hares often stand up on their hind legs and paw at
each other like boxers. This may be a struggle for
social dominance, or a Jill rebuffing a Jack.

Found in lowland grassland.
Widespread but declining.
Introduced into Ireland.

Brown hare *Lepus europaeus*

Its large eyes and long ears provide the hare with excellent
eyesight and hearing to give it plenty of warning of danger in the
open downs or farmland it inhabits. The best time to see one is in
the early morning or late evening as it sits up to survey its
surroundings. Even when feeding it rarely keeps its head down
for long. No other British mammal is better able to survive in a
totally open habitat, where cold, windy or rainy weather is as
much a challenge to survival as eluding predators.

Once brown hares were a common sight in pastures and
ploughed fields, but in recent years numbers have declined
markedly although they are still common in Scotland. The
reasons for the decline probably include intensified farming, with
larger fields. When crops are harvested the land is often
immediately ploughed, leaving a large area without food for
hares. Weedkillers deprive them of important food plants, while
pesticide sprays may contaminate food and kill leverets. The
young may also fall victim when crops are mown for silage.

A hare may live for three or four years – some may be twice
that age. Predators such as foxes, owls and buzzards take a few
leverets, but adults are fast enough to elude most enemies.

Summer
(June–Sept.)

Spring
(Feb.–May)

Winter
(Oct.–Feb.)

Mountain hares moult three times a year. In winter their fur turns from brown to white or skewbald. In spring they turn brown again, and in summer they grow a thinner, shorter, grey-brown coat.

In Ireland, mountain hares usually stay brown all year. They are often seen in large groups on farmland.

In summer, the mountain hare supplements its heather diet with rushes, cotton grass and bilberry shoots. It regularly eats its own soft droppings.

Eagles and harriers kill mountain hares, especially leverets (young). The hares may also fall prey to buzzards or wild cats, but foxes are the main predators.

Regular pathways are made across the moors where mountain hares nibble their way through heather and bilberry scrub.

The mountain hare turns wholly or partially white in winter. It is smaller than the brown hare, with a greyer coat, and its ears are also shorter – about 3 In. (75 mm) long. The tail is all white. Usually hares are seen alone or in pairs. About 20 in. (50 cm) long.

Patchy winter coat

A short burrow or a hollow scraped in snow may shelter the mountain hare from the cold and wind. Many starve when crusted snow prevents foraging.

Common on upland heather moors, but found at all altitudes in Ireland.

Mountain hare *Lepus timidus*

High heather moors are the chief home of the mountain hare, and heather is its main food. Because of the blue-grey tinge to its summer coat, it is also known as the blue hare. In summer its all-white tail helps to distinguish it from the rabbit and the brown hare, both of which have dark tops to their tails. The mountain hare's dense white winter coat is more likely to turn wholly white in colder temperatures or at higher altitudes.

The hares feed mostly at night, resting during the day in a scraped hollow or a form nibbled out among the heather. On mountain-sides they tend to move downhill to feed, returning to high ground to rest. In winter they move down from the highest areas to shelter in hollows or hide among boulders, and some 20 or more animals may gather in the same area. Mountain hares breed between February and August and may have three litters a year, each usually with one or two leverets. In some years more than three-quarters of the population dies before the next breeding season. Some may live to be ten years old.

In Ireland, mountain hares replace brown hares, living in all sorts of country including lowland pastures, and are often known as Irish hares. They used to be depicted on Irish coins.

169

Red deer retreat from high ground for the winter and live in the valleys. When spring comes they move back up the mountainsides to escape disturbance from people and pestilential flies.

Where there are meadow pipits there may also be pygmy shrews. Both feed on tiny insects among the stones and heather. Pygmy shrews keep well hidden but may be heard squeaking.

The stoat's white winter coat is good camouflage when hunting. It may also help to conserve body heat, because white fur radiates less than dark fur.

Many upland areas now have commercial plantations of dense conifers. These rob mountain hares of living space but may provide a home for pine martens.

Small herds of feral goats may be seen in a few mountainous areas. They vary widely in colour, ranging from white to piebald, skewbald, dark brown or black. Young ones are not likely to be seen before April.

In their winter coats, mountain hares look like patches of snow against the dark heather. They show up only when they move. On upland slopes they take the place of rabbits and brown hares.

Mammals of the mountain-sides

Britain's mountains are not very high – only seven are above 4,000 ft (1,220 m) – but they are bleak places for warm-blooded creatures to live, especially in winter. For roughly every 500 ft (150 m) of height above sea level, it gets almost 2°F (1°C) colder. Rain and mist as well as stronger winds also add to the chill. Trees rarely grow on high, windswept uplands, so there is little shelter. Consequently, few mammals live on the exposed high ground.

Although they forage on the heights in summer, red deer and, to a lesser degree, mountain hares move to lower ground for the winter. Small mammals, such as the field voles that live amid the upland grass, do not make long migrations. Although they lose a lot of body heat because they have a large surface area of skin in relation to their bulk, they can find shelter in burrows or among thickets of heather and bilberry. Despite the cold, small spiders and insects are surprisingly numerous amid upland vegetation, and provide food for pygmy shrews. Stoats thrive on field voles but will also take meadow pipits and other upland birds. Crows and foxes search the mountain-sides for carrion such as dead hares and sheep.

171

Anthills make good look-out points. Many have rabbit droppings on top, being used as rabbit latrines and to mark the limit of a territory.

Rabbits keep their fur clean by regular washing and grooming, using tongue, teeth and claws.

Shallow scrapes in the turf, exposing patches of soil, are common around rabbit warrens. They are made as territory markers.

When danger threatens, a rabbit warns others in the colony by thumping with the hind foot. A flash of white from the underside of the tail (scut) as it runs away also serves as an alarm signal.

A rabbit may sometimes be seen 'chinning' – rubbing the ground with its chin, which has scent-secreting glands. This is done to mark the rabbit's territory.

Orange nape

Black rabbits are not uncommon, particularly on offshore islands where populations are isolated. Albinos are rare, except on some offshore islands.

Grey-brown fur

Short tail

Rabbits generally have greyish-brown fur with orange at the nape, and a short tail that is black on top and white underneath. They are smaller than hares and have shorter ears without big black tips. Male (buck) 19 in. (48 cm) long. Female (doe) smaller with a narrower head.

Common. Increasing in many areas. Found mainly in farmland, lowland forests.

Rabbit *Oryctolagus cuniculus*

It was not until the 12th century that there were rabbits in Britain. They were introduced from the Continent as a valuable source of meat and skins. For long they remained a profitable part of the rural economy, but in the last 200 years have become major agricultural pests.

Rabbits are sociable creatures and live in colonies in burrow systems known as warrens, where a status system rules. Dominant rabbits, either male or female, claim and mark a territory in the best part of the warren and are more successful at breeding than subordinate rabbits. A subordinate does not establish a territory, and mixes amicably with other subordinates. But if it enters the territory of a dominant individual, it is driven off.

Most rabbit activity takes place at night and close to the warren. The animals rarely move more than 150 yds (about 140 metres) away from home, so the vegetation near the warren is kept short by frequent grazing. This leaves a wide, open area for revealing the approach of predators – rabbits have a good sense of smell and hearing and their prominent eyes are set so that they can see all ways at once. Short vegetation also helps the rabbits to keep their fur dry.

The breeding nest is in a dead-end burrow, or stop, often separate from the main warren. The doe lines it with fur plucked from her chest. Fur is loosened by hormone action.

Young rabbits leave the nest when they are about three weeks old, and are weaned about a week after that.

Rabbit families and feeding habits

In one year a female rabbit can produce more than 20 offspring, many of which will themselves breed when only four months old. Such prolific breeding is countered by deaths from cold, wet, disease and predators. Well over three-quarters of all rabbits live less than one year. The introduction of the flea-borne virus disease myxomatosis, in 1954, caused the deaths of more than 95 per cent of Britain's rabbits, and also brought about a decline in the numbers of buzzards, foxes, stoats and other predators that fed mainly on rabbits. Now rabbits are once more common, having developed a resistance to the virus, and are again causing damage to crops.

Rabbit warrens used to be sited in open fields but, because frequent cultivation and deep ploughing have made fields less suitable, are now more often in hedgerows or at field edges. From here the rabbits venture out to nibble plants and gnaw tree bark, mostly under cover of darkness. While underground during the day, a rabbit eats the soft droppings of food that has already passed once through its digestive tract, thus absorbing the vitamins and proteins still in the droppings. The hard, dark, second droppings are deposited above ground.

The breeding season extends from about January to August. A litter usually numbers from three to seven young. Their eyes open when they are about a week old.

174

Trees are damaged by rabbits eating bark and will die if the bark is eaten all round the trunk. Plantations often have rabbit-proof fences, or young trees are fitted with protective 'collars'.

Green plants of many kinds are eaten, nibbled off close to the ground as the rabbit works round in a semi-circle. Rabbits do occasionally drink, but usually get sufficient moisture from their intake of green food.

A layer of snow can prevent rabbits grazing at ground level but enables them to nibble snow-laden branches that are normally out of reach.

175

Nettles and elder bushes thrive near a warren, where the soil has been enriched with nitrogen from rabbit droppings.

Rabbit-proof fencing extends 12 in. (30 cm) underground. It protects sprouting corn, carrots, turnips and other crops from nibbling rabbits.

Short turf favours orchid growth. It also encourages insects because it warms quickly in the sun, faster than ground under long grass.

The warren (burrow system) is dug in deep soil and has many entrances. Birds such as puffins and jackdaws may take over old cliff-top burrows.

For buzzards, rabbits are a major food. As with foxes and stoats, their chances of survival and successful breeding increase with an abundance of rabbits.

How rabbits transform the landscape

Without rabbits, much more of our downland and cliff tops would be a mass of bramble and hawthorn scrub – not the short, flower-studded turf that is so characteristic. Rabbits suppress the growth of shrubs by nibbling the growing shoots, but they tend not to eat older shrubs, so established thickets remain. These provide rabbits with shelter from hungry buzzards, but give cover to hunting foxes. Nibbling rabbits are very damaging to crops, so farmers try to fence off their fields. Cliff tops are not suitable for crops because of exposure to wind and salt spray, so rabbits are usually tolerated along coastal strips.

By their close and constant grazing, rabbits crop the grass as short as if it had been mown. Short turf favours the growth of low-growing or creeping plants such as vetches and trefoils, which would otherwise be swamped by long grass and shrubs. These plants attract many butterflies, such as the common blue. The butterflies feed on nectar and lay their eggs on the plants, which serve as food for the caterpillars. The short turf is also highly suitable for other insects, especially ants. The insects in turn attract many species of birds – skylarks for example, common in open country.

Rabbits feed mostly at night, but in undisturbed places will graze by day. Turf near a burrow may be cropped very short, but thistles are avoided.

HOW HORSESHOE AND ORDINARY BATS DIFFER

There are 19 families of bats throughout the world, comprising 950 different species. Only two of the families are found in Britain, the *Rhinolophidae*, or horseshoe bats, and the *Vespertilionidae*, or ordinary bats. There are two British species of horseshoe bats and about a dozen of ordinary bats.

Many people dislike bats, but they are harmless and very interesting creatures. They are the only mammals capable of flying long distances, using a modified form of the limbs common to all mammals. Scientists classify bats as the Order *Chiroptera*, which means hand–wing. Bats do not walk far and have poorly developed hind limbs.

When flying fast and in the dark, bats can still avoid obstacles and track down prey by means of elaborate echo-location systems similar to the radar scanning equipment used in ships and aircraft. The two British bat families have different echo-location systems. Horseshoe bats emit sound through the nostrils and it is focused into a narrow beam by a fleshy, cone-shaped trumpet on the snout. The bat moves its head from side to side to scan ahead. Ordinary bats emit sound through the mouth, and have in each ear a fleshy spike known as a tragus, which is part of their sound-reception system.

All British bats eat insects, and as these are scarce in winter the bats hibernate during the coldest months. Many bats make seasonal migrations to hibernating places, and some foreign bats turn up on North Sea oil platforms or ships, having been blown off course.

FACIAL DIFFERENCES

No fleshy spike in ear

Greater horseshoe

Nose-leaf

Horseshoe bat

The bat has a horseshoe-shaped fold of skin round the nostrils, with a triangular lobe projecting upwards between the eyes. The ears lack a fleshy central spike or lobe, known as a tragus.

Fleshy spike in ear

Dog-like muzzle

Serotine

Ordinary bat

The bat has a dog-like muzzle with no nose-leaf. But it has a fleshy spike (or central lobe), known as a tragus, in each ear. The shape of the tragus varies with each species of bat.

HOW ALL BATS ARE BUILT

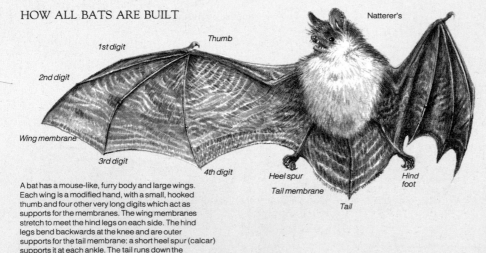

Thumb
1st digit
2nd digit
Wing membrane
3rd digit
4th digit
Natterer's
Heel spur
Tail membrane
Tail
Hind foot

A bat has a mouse-like, furry body and large wings. Each wing is a modified hand, with a small, hooked thumb and four other very long digits which act as supports for the membranes. The wing membranes stretch to meet the hind legs on each side. The hind legs bend backwards at the knee and are outer supports for the tail membrane; a short heel spur (calcar) supports it at each ankle. The tail runs down the middle of the membrane.

WINGS AND TAILS COMPARED

Greater
horseshoe

Horseshoe bat
The wings are broad and rounded. They allow slow, unhurried, rather than fluttery flight, and easy turning in small spaces such as caves. The tail is short and the membrane shallow.

Ordinary bat
The wings are narrower and more pointed at the tips. Some species, such as the noctule, have particularly long, narrow wings that allow fast flight but less ability to manoeuvre in small spaces. The tail is usually longer than that of a horseshoe bat, with a deeper membrane, but the pattern varies according to the species.

DIFFERENT WAYS OF RESTING

Horseshoe bat
The bat always rests by hanging upside down and gripping with its toes. It sleeps with its wings wrapped round it like a shawl. Because of its rounded body it finds crawling difficult, so needs a roost to which it can fly direct.

Greater
horseshoe

Ordinary bat
The bat roosts by hanging, head down, generally with its wings folded at its side. Often, it squeezes into a crevice in bricks or wood. It has a flat body and can crawl with its wings folded, using its forelimbs as legs and its hooked thumbs to get a grip on the surface.

Noctule

Pipistrelle

179

Horseshoe bats need roosts where they can hang freely. They do not hide in crevices like other kinds of bat. Lesser horseshoe bats often use cellar ceilings, or may hang from a protruding nail.

A horseshoe bat's nose-leaf is used to direct the high-pitched sounds it emits to navigate or find its prey, locating it by the echoes. The lesser horseshoe has the highest-frequency sound pulses of any British bat, which help it to locate very tiny prey.

Its broad, rounded wings enable a horseshoe bat to fly slowly, hover or fly in a narrow space such as a chimney. Lesser horseshoe bats eat tiny insects such as gnats, caught in their jaws in flight.

Like birds, bats can be fitted with identity tags (on the wrist) to trace movements and life-span. Tags have revealed that a lesser horseshoe bat can survive for more than 15 years, but most live only four or five years.

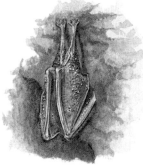

A hibernating horseshoe bat wraps its wings round its body. Because it hibernates in a humid place, it is often covered with tiny droplets of condensation.

Lesser horseshoe bats are most likely to be seen hanging singly or in small groups in a cool, damp place such as a cellar. They fly after sunset.

Broad wings

Found where there are caves or mines for hibernation. A few recorded from east England.

The lesser horseshoe bat has a body smaller than a man's thumb, and is one of Britain's tiniest mammals. Like the greater horseshoe bat, it has a nose-leaf (horseshoe-shaped skin round the nostrils), a rounded body and broad wings. Female 1½ in. (40 mm) head and body, 10 in. (25 cm) wingspan. Male smaller.

Lesser horseshoe bat *Rhinolophus hipposideros*

The lesser horseshoe bat is a delicate, very much smaller edition of the greater horseshoe bat, and more often solitary. Like its large, robust cousin, the bat likes to hibernate in damp cellars, mines or tunnels in winter, so lives only where such places are available. Because of its smaller size, however, its choice is less restricted, for it can fly in a tunnel only 6 in. (15 cm) high, or up a shaft only 20 in. (50 cm) wide. In Britain lesser horseshoe bats do not seem to travel very far from their regular haunts, but on the Continent journeys of up to 95 miles (150 km) have been recorded.

The bats mate in autumn and winter, but the embryo does not begin to develop until April. A female produces only one baby a year, born usually in July or August in an attic, a hollow tree or an old building. Youngsters can fly around the roost when three weeks old, and all can look after themselves by late August. Young bats do not normally breed until their second year. Although a lesser horseshoe bat is similar in size to a common shrew, its life-span is probably four or five times longer. Winter hibernation may save the bat from some of the wear and tear experienced by the ever-active shrew.

181

Mating takes place in autumn and winter, but fertilisation does not occur until spring. Young are born mostly in July.

Females and babies congregate in a nursery roost, often in a roof space. Females breed from their third year and produce one baby a year. Young bats fly at about three weeks old and leave the roost at about five weeks old.

The bat emerges after sunset to hunt, often flying low in search of insects such as dung flies. It emits high-pitched sounds and locates prey by their echo, carefully patrolling a good place before moving on.

A baby can be carried by its mother, clinging to the false teat near her hip, until it gets too big for her to manage.

The bat cannot crawl. It flies straight to a roost and hangs by gripping the rough surface with its claws. A bat's weight keeps its toes gripped; it bends its knees to release itself. It may bend them if disturbed.

Large prey such as a cockchafer beetle may be taken from near the ground, the bat swooping low to grab the creature in its jaws.

The bats hibernate in clusters in humid caves, but are now so rare that few large clusters are seen. Clustering helps to maintain a suitable temperature.

Broad wings

Its large size and broad, rounded wings distinguish the greater horseshoe bat. It gets its name from the horseshoe-shaped skin (nose-leaf) – used for echo-location – round its nostrils. Female 2½ in. (64 mm) head and body, 13½ in. (34 cm) wingspan. Male smaller.

Declining. Found only where suitable hibernation places and conditions still exist.

Greater horseshoe bat *Rhinolophus ferrumequinum*

Once common in southern Britain, greater horseshoe bats are now scarce and in danger of extinction. Like other bats, they have suffered considerably from the decline in the number of places where insects abound – such as hedges, ponds and old grassland – as well as from eating insects contaminated with pesticides. Another handicap is that southern Britain is the extreme northern limit of the greater horseshoe bats' range. They really prefer a warmer climate. If there are several cold or wet summers in succession, the bats do not breed very successfully and the young have difficulty in surviving their first year.

Greater horseshoe bats also need to find somewhere for winter hibernation – large, spacious, humid places such as old mines, caves or damp tunnels – and their numbers and distribution are limited by the availability of such places. Even where suitable places exist they may be frequently disturbed or be blocked up for safety. Efforts have been made to protect the bats' favourite breeding and hibernation sites and, like other British bats, they are protected by law. But these measures may not be enough to save a species whose numbers have shrunk to probably less than a quarter of what they were 20 years ago.

183

IDENTIFYING ORDINARY BATS

There are about a dozen species of ordinary bats (bats of the *Vespertilionidae* family) that may be seen in Britain. They cannot be identified with confidence unless they are closely examined in the hand. Face colour, tail structure and wingspan are the main aids to recognition, also the size, shape and colour of the ears and the size and shape of the tragus – a small, upright projection of skin in the ear. The tragus may be broad, narrow, short or more than three-quarters the length of the ear itself.

BATS WITH DARK FACES

Noctule
Nyctalus noctula
Page 188

Muzzle very broad, bare. Ears rounded with short, semicircular tragus.

Each tail membrane has a tiny lobe on the edge, just behind the heel spur (calcar). Leisler's similar but smaller and darker.

Large: wingspan 14 in. (36 cm).

Pipistrelle
Pipistrellus pipistrellus
Pages 186–7

Muzzle broad with snub nose. Ears short with a short tragus.

Each tail membrane has a tiny lobe on the edge, just behind heel spur (calcar).

Tiny: wingspan up to 8½ in. (22 cm).

Whiskered
Myotis mystacinus
Page 198

Small: wingspan 9¼ in. (24 cm).

Tail membranes are straight-edged. Feet tiny with toes almost parallel.

Muzzle narrow. Ears dark, narrow with a long, pointed tragus.

Barbastelle
Barbastella barbastellus
Page 194

Muzzle flat. Ears black, broad and squarish, meeting between eyes. Tragus tall, spear-like.

Serotine
Eptesicus serotinus
Pages 190–1

Large: wingspan 14 in. (36 cm).

Tail projects beyond edge of tail membranes.

Muzzle broad. Ears dark with short, pointed tragus.

Medium: wingspan 10¼ in. (26 cm).

BATS WITH LIGHT BROWN OR PINKISH FACES

Daubenton's
Myotis daubentoni
Page 199

Muzzle broad, blunt.
Ears short, brown, with
narrow, pointed tragus
about half ear length.

Medium: wingspan
10 in. (25 cm).

Each tail membrane
joins leg half-way
up shin. Feet large
with spreading
toes.

Long-eared
Plecotus auritus
Pages 192–3

Muzzle rounded. Ears very long,
oval, joined at forehead. Tragus
long, narrow, pointed. Grey
species similar except for colouring.

Medium: wingspan 10 in. (25 cm).

Mouse-eared
Myotis myotis
Page 196

Muzzle long. Ears long, brownish,
separate at forehead. Tragus
long, pointed.

Very large: wingspan
16 in. (40 cm).

Natterer's
Myotis nattereri
Page 195

Muzzle fairly long,
narrow. Ears long,
drooping, each with
long, pointed tragus.
White belly.

Medium: wingspan
11 in. (28 cm).

Tail membranes baggy
with outward-curving
edges and fringe of very
fine hairs.

Bechstein's
Myotis bechsteini
Page 197

Muzzle fairly long and narrow. Ears long,
separate at forehead; droop sideways
at rest. Tragus long and narrow.

Medium: wingspan
11 in. (28 cm).

185

Young are born in a
large nursery colony,
which is often in a roof
space behind tiles.
They fly after about
three weeks.

In summer the bats like to
squeeze into warm crevices
in house roofs or behind wall
tiles. Some live in hollow trees
or behind bark.

Flight is fast and fluttery with
frequent twists and spiral
dives as the bats chase prey.
They feed mostly on small caddis-
flies, gnats and tiny moths
and are often seen flying
low near water.

Small insects are
eaten in flight. Larger
ones may be taken to a
feeding roost such as
a tree branch. Food
remains accumulate
on the ground below.

A mother and baby can fit on a
standard-size brick with plenty of room
to spare. Young bats have darker fur
than adults.

Short,
broad ears

The bat clings to a roughened surface with its hind feet as it hangs head down ready to launch itself into flight. It can crawl upwards by using its hooked thumbs.

Pipistrelle bat *Pipistrellus pipistrellus*

The tiny pipistrelle is Britain's smallest bat and also the most abundant and widespread. Pipistrelles congregate in colonies, often in buildings such as churches, and can be heard squeaking before they stream out 15-30 minutes before sunset to forage for insects. Colonies may number 200 or more bats. Favourite summer roosting places are small, warm spaces behind tiles on a south-facing tile-hung wall, or behind weather-boarding or wooden shingles. Such features are found on many fairly modern houses and bungalows, so pipistrelles may be common in areas with new housing.

Warmth is essential for the tiny young, born mostly in June. They are hairless for a week after birth and liable to become chilled. The warmer they are, the faster they can grow. In summer, mothers and young live in separate colonies from the males. Winter colonies contain both sexes and it is during winter that mating occurs, although fertilisation is delayed until about April. Pipistrelles hibernate from about late November until late March, preferring cool, dry places such as churches, house roofs or old trees. They fly out sometimes during hibernation, occasionally by day.

Common. Widespread in all habitats, cities included. May be rare on high ground.

Pipistrelle bats are very tiny, with short, broad ears and fairly narrow wings. Adults vary in colour from place to place; in some colonies they are mainly orange-brown, in others pale grey-brown. The bats are common in buildings. About 1⅜ in. (35 mm) head and body, 8½ in. (22 cm) wingspan.

The golden-brown hairs of the noctule bat's fur are a uniform colour from the base to the tip.

Short, rounded ears

Dark face

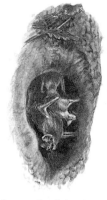

Noctule bats often live in hollow trees or woodpecker holes. They have short, rounded ears with a rounded tragus (the central lobe in the ear) and a dark brown face. About 3 in. (75 mm) head and body, 14 in. (36 cm) wingspan.

Common in southern Britain, especially in the Midlands and south-eastern counties.

Prey is mostly larger insects caught on the wing. The bat will swoop low to catch an insect such as a cricket.

Young are born in June or July. Usually there is only one baby, but females may sometimes have twins.

The noctule has a broad muzzle and many face glands that give a strong smell to both the bat and its roost.

Noctule bat *Nyctalus noctula*

The noctule bat is a powerful and expert flyer, so it is strange that, although common in much of England and Wales, it is not found in Ireland or most of Scotland. It is one of the largest British bats and could easily fly across the Irish Sea. On the Continent, noctule bats regularly migrate hundreds of miles.

High-flying noctule bats are often on the wing before dark in summer, swooping after insects among swifts and martins. They are mostly tree dwellers, roosting in colonies in tree-holes in parks and woods, although they will sometimes use a house roof. If a roost becomes too hot or too crowded, they move to another. Noctules may have difficulty in finding roosts, because nesting starlings take over many suitable tree-holes.

Like most bats, noctules form nursery roosts of females and young in summer. The young are born in June or July and weaned at about a month old. In winter, from about October until the end of March, the bats hibernate in trees and buildings. They do not use caves, where temperatures are more stable, which suggests that they can stand low temperatures better than most other bats. Noctules probably live for ten years or more, and have few predators.

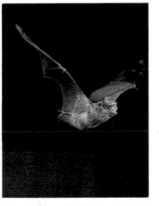

Like all Vespertilionid bats, Leisler's locate prey with echoes of high-pitched sounds emitted from their mouths.

Short, rounded ears

Broad muzzle

In summer adult males gather in small groups and live apart from the females and young.

Patchily distributed. May be uncommon. In Ireland it replaces the noctule bat.

Leisler's bat is very similar to the noctule bat, with short, rounded ears and a broad muzzle. Although it is smaller than the noctule and has slightly darker fur with paler tips, it is hard to tell the two bats apart. About 2½ in. (64 mm) head and body, 12 in. (30 cm) wingspan.

Leisler's bat *Nyctalus leisleri*

Leisler's bat is also known as the lesser noctule bat, and is very similar to the noctule. It differs only in its smaller size and the fact that its glossy bronze fur is paler at the tip rather than a uniform colour throughout. Both species are high flyers, often about before sunset, but have a curiously different distribution. Leisler's bat is present in Ireland, where there are apparently no noctules. But in Great Britain it has been recorded only rarely, and then in widely separated localities.

Like the noctule, Leisler's bat has a short, very broad muzzle and a semi-circular tragus (central ear lobe) – features not found in other species of British bats. It is a tree dweller, roosting in tree-holes in summer and hibernating in them in winter – from about October until the end of March. Nesting starlings probably deprive Leisler's bats of many suitable roosting holes, added to which they also have to compete for them with the larger noctules. Little is known about the Leisler's breeding habits but, like noctules, the females probably each produce one baby a year in June or July, rearing them in nursery roosts. Leisler's bats have few creatures preying on them and their normal life-span is probably ten years or more.

The bats emerge shortly after sunset and are high flyers. They are large bats, but seem even bigger and heavier in the failing light.

Young are born singly. A mother can carry a small, pink baby clinging to her body if the nursery roost is disturbed. An adult's tail tip projects noticeably beyond the edge of the tail membrane.

The gable end of a brick house with a slate roof is a typical site for a nursery colony, usually numbering 50–100 bats.

The bat crawls with fore-wings folded, hooking its thumbs on a rough surface while it pushes with its hind feet.

Slow-flying serotine bats eat mostly large moths and beetles, which are caught on the wing.

Most serotine bats are born in late June and are blind and hairless at birth. They are able to fly by the time they are about four weeks old.

Grizzled fur

Dark face and ears

The serotine is a large bat with powerful jaws, a dark face and ears and a pointed tragus (the central lobe in the ear). Its brown fur is paler at the tips, making it look grizzled. About 2½ in. (64 mm) head and body, 14 in. (36 cm) wingspan.

Common in parts of southern and eastern England. A few isolated sightings in north.

Serotine bat *Eptesicus serotinus*

Although it is a large bat and a strong flyer, the serotine has a surprisingly localised distribution. It is common only in certain areas of southern and eastern England, especially in the south-east, and where it occurs, it is the large bat most often seen. There are hardly any records of serotine bats being seen to the north or west of the Midlands.

Serotines particularly like to raise their young in the attics of old houses, often hanging up along the roof ridge. Unlike the tiny pipistrelle bat, they are too large to squeeze behind slates and tiles. Colonies may return to the same place year after year, causing an accumulation of small, very black droppings in the roof space. These differ from rat or mouse droppings in containing shiny pieces of chewed insects, and although they can be smelly they are not a health hazard. Nursing colonies can be quite noisy and may be unpopular with householders, but it is illegal to disturb them. In any case, once the young are able to look after themselves – usually about August – the colony departs. They spend the winter elsewhere, hibernating in another roof or a hollow tree. Serotines probably live for about five years. Some may survive to be 15 or more.

191

To reduce moisture loss while hibernating, the ears are folded back alongside the body and tucked under the wings. The tragus, a spear-like central lobe in the ear, remains hanging down.

Unlike other bats, the long-eared bat feeds extensively on insects or larvae that are resting on foliage. It hovers while it seizes prey.

Grey long-eared bat

Plecotus austriacus

It is very difficult to tell the difference between the grey and common long-eared bats. The grey is slightly larger and has greyer, darker fur. It is rare in Britain, but widespread on the Continent. A few are found in the south in Dorset and Hampshire.

Fine struts of cartilage run across the ears to help to hold them erect. The ears concertina at the outer edge to droop when the bat rests.

The bat flies with ears erect. Its broad, rounded wings allow slow flight and easy movement in small spaces. It can hover with its body almost upright.

Huge
ears

Nursery colonies of females and their young often cluster in attics. Females produce one baby each in late June or July. The young fly when a few weeks old.

Common long-eared bat *Plecotus auritus*

All British bats use their hearing to navigate and to locate food. They emit sound – too high-pitched for human ears – that bounces back from obstacles, including tiny insects. From these echoes a bat learns the distance and direction of an object. The long-eared bat's huge ears are part of a system of echo-location sensitive enough not only to detect flying insects, but also to distinguish between an insect or larva and the leaf it is on.

The long-eared bat can also fly with fine control in small spaces. This, together with its refined echo-location, enables it to thread its way through tree branches and foliage and even hover above a leaf to pick off an insect. It is principally a woodland bat that roosts in trees, although it often breeds in attics. Long-eared bats hibernate, usually alone, from about November to March. Sometimes they hibernate in the summer roost, which is unusual among bats, but commonly seek out a different place to spend the winter, such as a cave or mine. Some live for 12 years or more. The common long-eared bat differs from the rare grey long-eared bat in colour. It has yellowish-tipped pale brown fur, whereas the grey has darker fur with hairs dark grey at the base.

Found in woods, house roofs. Widespread, but not in open, exposed places or far north.

Huge oval ears meeting at the base distinguish the long-eared bat. The ears are about 1⅛ in. (28 mm) long, nearly three-quarters of the length of the head and body. The bat has broad wings and yellowy or light brown fur, paler below. About 1¾ in. (45 mm) head and body, 10 in. (25 cm) wingspan.

193

Although a barbastelle usually alights head up, it may twist round before attaching itself to hang head down from a wall or tree-trunk.

Frosted coat

Pug-like face

Short, broad ears

The distinctive barbastelle bat has a pug-like face with a bare, dark brown snout and short, broad ears that meet between the eyes. The glossy fur is almost black, and on older bats is cream tipped, giving them a frosted look. About 1¾ in. (45 mm) head and body. 10½ in. (27 cm) wingspan.

Barbastelle bats are found in open wood-land. They usually live in trees, occasionally in houses.

Barbastelle bat *Barbastella barbastellus*

No other British bat bears any resemblance to the strange-looking barbastelle, with its squashed face, thick black ears that meet between the eyes and long, frosted-looking, blackish fur. It is so peculiar that there is only one other species like it in the world, another barbastelle found in the Middle East and Asia.

Barbastelles seem to be uncommon, because despite their distinctive looks, sightings are rarely reported. They are found in open woodland, especially in river valleys, where they fly low over the water and are active intermittently through the night. As with all bats, the males are smaller than the females and tend to be solitary, leaving the females to form nursery colonies in summer to raise their young. In winter, barbastelles have been found hibernating in caves and ornamental grottoes with other bats, but only in the coldest weather. They tend to choose the coolest places to hibernate, which suggests that they are hardy and able to tolerate low temperatures so can normally spend the winter in relatively unsheltered sites. Although in Britain the barbastelle has been recorded mainly in the south, on the Continent it is found as far north as latitude 60° and eastwards into Russia. It can live for at least 18 years.

Found only in the southern half of Great Britain, but elusive and rarely seen.

The ears are almost as broad as they are long, and have a stiffening fold at the front. There is a long, pointed tragus (central ear lobe).

The ears droop sideways and have distinctive upturned tips, which are darker than the pinkish bases. If laid forwards they extend well beyond the bat's nostrils.

In winter, Natterer's bats will often hibernate in caves. Mating occurs during wakeful periods at this time.

White underside

Natterer's bat has a distinctive pure white underside, and its tail membranes are baggy with outward-curving edges. Its ears are drooping and rather long, and its face long and reddish with a bare, narrow muzzle. About 1¾ in. (45 mm) head and body, 11 in. (28 cm) wingspan.

Common. Widespread in woods, farmland and parkland. Not found in northern Scotland.

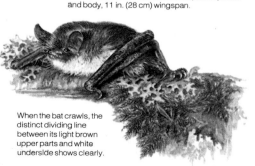

When the bat crawls, the distinct dividing line between its light brown upper parts and white underside shows clearly.

Natterer's bat *Myotis nattereri*

Natterer's bat is named after its discoverer, an early 19th-century Austrian naturalist. It is most easily recognised by its pure white underside as it flies slowly and often at rooftop height soon after sunset, searching for small flying insects such as moths. It will sometimes pick insects off foliage. In flight the bat often holds its tail pointing downwards instead of trailing to the rear, as is usual in other bats. Along the edge of each tail membrane there are about ten tiny, bead-like swellings, each sprouting a short, very fine hair that is scarcely visible. The function of this fringe is unknown; Natterer's is the only bat species with such a feature.

In summer, breeding colonies of females and young gather in hollow trees or house roofs. As with many other bats, the females bear only one baby in June or early July. It can fly by about August. During their first year, young Natterer's bats can be distinguished from adults by their colouring, being greyish-brown, including their underparts. The bats do not generally hibernate until December, emerging in early March. They hibernate in caves if these are available, but seem content to use hollow trees or other sites. They can live for 25 years.

195

Wide wingspan

A nursery roost
may be in a warm
dry attic or an old
railway tunnel.

Mouse-eared bats hibernate in cool
places, usually alone. A hibernating bat
may be covered with condensation.

The entrances to old limestone
mines used for hibernation have
been fitted with grilles to prevent
bats being disturbed.

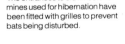

Some colonies recorded in
Dorset and Sussex, but
probably now extinct in Britain.

Large ears

Britain's biggest bat, the mouse-eared bat is
distinguished by its wide wingspan, bare,
pinkish face and large, wide ears. Its grey-
brown back and greyish-white underparts are
distinctly divided. About 2¾ in. (70 mm) head
and body, 16 in. (40 cm) wingspan.

When the bat crawls, the sharp
dividing line between its body
colours shows clearly.

Mouse-eared bat *Myotis myotis*

Its slow, heavy flight and large wingspan – almost one-third
larger than that of any other British species – easily distinguishes
the mouse-eared bat. It is the biggest bat known in this
country, but is probably now extinct. Because of its scarcity it has
had legal protection since 1975.

On the Continent, mouse-eared bats are fairly common in
places, and sometimes form large colonies. In Britain the only
known regular haunts have been in Dorset and Sussex, and it is
possible that these bats flew across the Channel from France,
although no one has proved that such flights took place. For a
mouse-eared bat seeking a suitable place to spend the winter, the
short flight from France to England might have been well worth
the effort. In eastern Europe mouse-eared bats are known to
travel 125 miles (200 km) or more between their summer and
winter quarters. Mouse-eared bats seem to prefer open woodland or farming country. In summer the females give birth to
their single young in a warm, dry place such as an attic, and the
males live alone elsewhere. The winter is often spent in the cool,
sheltered conditions of a cave or an old mine tunnel. The bats can
live for 15 years or more.

The rare Bechstein's bat may use an attic as a nursery roost and has also been known to roost in a nest-box.

The long ears droop to the side when the bat is at rest. The ears do not meet at their bases like a long-eared bat's.

Long ears

When the bat roosts, its ears hang down. They are not folded up under the wings.

Narrow muzzle

Found only in the extreme south of England, but even there very rarely seen.

Next to the long-eared bat, the rare Bechstein's bat has the longest ears of any bat in Europe – they measure about 1 in. (25 mm). Its muzzle is pinkish-brown and fairly long and narrow. About 1¾ in. (45 mm) head and body, 11 in. (28 cm) wingspan.

Bechstein's bat *Myotis bechsteini*

The elusive Bechstein's bat is one of Britain's rarest mammals. As with the mouse-eared bat, only a few dozen sightings have ever been made. Most of these have been in Dorset, where it hibernates in old limestone mines along with several other kinds of bat. Its distinctive, long-eared appearance makes Bechstein's bat easy to distinguish from other species, so the lack of sightings probably reflects genuine scarcity. The bat ranges throughout Europe but is very scarce there too. Most sightings have been in Germany, and the bat is named after an early 19th-century German naturalist. It was first discovered on the Continent and only later found to occur in Britain.

Bechstein's bat seems to be a woodland species that frequents parkland, often flying about 10 ft (3 m) above the ground, and roosts in hollow trees. Its bones have been found in old flint mines that date back to the Stone Age, when most of southern Britain was covered by extensive forests. Like other bats, female Bechstein's give birth to one baby a year. The nursery roost is often in an attic or tree-hole. The bat probably feeds mostly on moths, sometimes other insects, taking them mainly in flight but occasionally off a leaf surface.

Narrow,
pointed
wings

Tiny feet

The bat may hibernate in a cellar, cave or fissured cliff. Buildings and trees are used as summer roosts.

The bat's pointed, upright ears give it a rather perky look. The tragus (central ear lobe) is narrow and pointed.

Widespread except in far north. Probably found in Ireland, but not recorded.

Whiskered bats are small and delicate, with narrow, pointed wings and tiny feet. The muzzle is narrow and the face and ears very dark, especially in young animals in their first year which also have darker underparts than adults. About 1½ in. (40 mm) head and body, 9½ in. (24 cm) wingspan.

Whiskered bat *Myotis mystacinus*

Despite its name, the whiskered bat does not have particularly prominent whiskers, but it does have more fur round the eyes and muzzle than most other bats. Common around buildings, hedges or woodland fringes, the bat may be seen from early evening – usually alone but occasionally in groups as it makes slow, fluttering flights in search of insects. Usually it follows a regular track repeatedly before moving elsewhere. It is active through the night but probably rests at times, hanging upside down against tree bark or some other rough surface.

Whiskered bats hibernate in cool places from late autumn. As with most other bats, mating takes place during short periods of wakefulness in winter. The females are therefore pregnant as soon as (or soon after) they come out of hibernation in early spring, allowing the single young to be born as early as possible in summer – usually about June. They then have plenty of time to feed, grow and fatten up before winter, when food supplies become insufficient for normal activity. Tiny whiskered bats can live for about 19 years. Although they do not seem to travel far in Britain, on the Continent marked bats have been known to journey some 1,200 miles (nearly 2,000 kilometres).

Nooks among masonry or rocks make daytime roosting places. The bats often return to rest during the night.

The bat's short ears are wider than those of the whiskered bat, and paler at the base.

Widespread in most parts of Britain except exposed places.

Daubenton's bat is often called the water bat because of its tendency to fly very low over water. It has short ears and a short, broad, pinkish-brown muzzle. The large feet with splayed toes are distinctive. About 1¾ in. (45 mm) head and body, 10 in. (25 cm) wingspan.

To hibernate, a bat may crawl backwards into a crevice in a cave, wall, rock-face or hollow tree.

Large feet

Daubenton's bat *Myotis daubentoni*

When it skims low over a pond or lake at night, Daubenton's bat uses shallow, fluttering wing-beats that enable it to get very close to the surface. In this way it can pick up mayflies as they emerge from the water, and may also manage to catch other aquatic insects, plankton and even small fish occasionally – a food source not exploited by other British bats. Daubenton's bat is widespread and usually found not far from trees or water. It does not always hunt low; it may be seen flying fairly high and it is most likely to be confused with the whiskered bat. At close quarters, however, its large feet distinguish it.

When hunting, a Daubenton's bat often patrols repeatedly up and down a regular beat, then moves to another. These favoured beats may actually be defended from other bats, but little is known of bats' territorial and social behaviour because of the difficulty of studying them in flight and in the dark. Like other bats, Daubenton's females bear one baby a year about June or July. Nursery roosts, with no adult males present, are often in buildings and sometimes contain scores of bats. During the winter, Daubenton's bats tend to be solitary, hibernating alone in caves or trees.

199

Caves provide sheltered, humid places in which bats can hibernate during winter. Horseshoe bats especially may travel long distances to find a suitable cave. Many caves are now blocked up.

Many kinds of bat like to roost or hibernate in old trees. As these fall or are cut down, bats have fewer places to live.

Many bats like to live in woodland. But conifer plantations, which have dense stands of trees of the same age, support fewer insects and lack holes where bats can roost.

Why bats are disappearing from Britain

Bats are probably the most endangered animals in Britain. All kinds are less common than they were about 30 years ago, and several species may be close to extinction here. The reason is that their two main needs – food and shelter – are both seriously threatened, especially by modern farming and forestry practices. Hollow trees and hedges provide bats with both shelter and food (insects), and many insects also breed in ponds. But the intense cultivation of arable crops leads to a 'tidy' countryside where hollow trees and hedges are removed, ponds filled in and wet areas drained. Crop spraying not only kills many insects, it contaminates the survivors; bats eating them accumulate poisons that may kill them or impair their breeding ability. Herbicides used to get rid of weeds destroy many plants that insect larvae feed on, so there are fewer insects for bats to eat.

Although bats are harmless and protected by law, ignorance and dislike causes many people to drive them out of places such as house lofts and old churches. But bats are useful because they eat many night-flying insects that damage crops and trees. They help to control insect numbers in a less destructive way than chemical sprays, which kill all insects (including plant pollinators) and often poison other animals as well.

Old grassland nurtures many insects for bats to feed on, and provides open country where they can fly and hunt. Bat numbers are dwindling as more grassland is ploughed.

All sorts of insect breed in ponds, particularly midges which are a major source of food for some bats. But ponds are often drained or filled in to make room for crops.

Hedgerows shelter a variety of insects that bats eat. But hedges are often removed to give farm machines more working space.

201

Frogs move by hopping or leaping; they do not crawl. Their long hind legs enable them to leap up to 20 in. (50 cm) from a standing start.

Frogs mature at about two years old and after that migrate early each spring to small ponds and ditches to breed. Continual croaking by males helps to attract others to the site.

Tadpoles hatch from the eggs after 14 days or more, depending on temperature. By about three months old they have developed four legs, ready for life on land.

When young frogs first leave the water, usually during June, they still have a stumpy tadpole tail, but this soon disappears.

The eggs (spawn) are laid in a mass in a transparent jelly that protects the embryo frogs and also keeps them slightly warm, so speeding development.

Frogs go to the same breeding place year after year. Because many natural ponds have become polluted or been drained, the commonest breeding places are now artificial garden ponds.

Dark patch
behind eye

Variable
skin colour

The common frog's smooth skin and long legs with bold markings distinguish it from a common toad. The frog also has more fully webbed hind feet.

Common frog *Rana temporaria*

Open woods and lush pastures are typically the home of the common frog, which likes moist places not too far from water. But frogs are becoming increasingly common in gardens with ponds, and it is not unusual to find a hundred or more using a pond for spawning. As frogs eat insects and small animals such as slugs and snails, they are good friends to the gardener. They like to hide in tall vegetation on summer days, and emerge on warm damp nights to hunt.

In winter, from about mid-October, common frogs hibernate in sheltered places on land or in the muddy bottoms of ponds. They emerge to migrate to breeding ponds, often in January in the south-west but usually in February or March in other parts. After spawning frogs usually stay in the water until the weather gets warmer, leaving during April to live on land. When the young frogs emerge in June or July, large numbers are killed by predators such as blackbirds. At any age they may fall prey to a host of animals, including crows, herons, ducks, hedgehogs, rats, foxes, grass snakes and cats. Frogs are now rare in farmland, mainly because of the increase in arable fields treated with pesticides, and the neglect or filling in of ponds.

The common frog lives on land in damp places for most of the year, often blending well with its surroundings. The dark patch behind its eye is distinctive, but body colour varies widely from dark greenish-grey to chestnut or yellow, or sometimes albino. Female about 3 in. (75 mm) head and body. Male slightly smaller.

Widespread in wet habitats, but land drainage is making it locally extinct.

Edible frog
Rana esculenta

Closely related to the marsh frog, the edible frog was introduced to Britain from the Continent in the 19th century. It is slightly bigger than the common frog, and its bright green colouring is distinctive. There are a few scattered colonies in the south-east and East Anglia.

Drainage dykes are the home of the marsh frog, which catches most of its food on land. The frogs breed in a dyke or at the edge of a large pool in late May or early June.

If disturbed as it sits on the bank, the frog dives into the water with a plop. Its strong hind legs enable it to clear 6 ft (1·8 m) in one leap.

In summer marsh frogs like to bask in the sun, on a bank or a floating piece of wood. But they keep alert for the approach of predators, such as a heron or a fox.

Tree frog
Hyla arborea

Suction pads on its toes enable the tiny, bright green tree frog to climb among trees and shrubs, where it catches flying insects at night. The frog was introduced from Europe earlier this century. There are colonies in the New Forest and in south-east London.

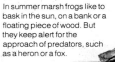

In April marsh frogs emerge from the water after hibernating in the mud or stones at the bottom of a pond. They do not breed until a few weeks later.

The marsh frog's larger size and more pointed snout help to distinguish it from the common frog. Its eyes are also closer together and have no black patch behind them. Males have two large vocal sacs that amplify their croaks at breeding time. Female 5 in. (12·5 cm) head and body. Male smaller.

Pointed snout

Vocal sacs inflated

Although it is bigger than a toad, a marsh frog has the smooth skin and leaping gait of a frog. Its colouring ranges from brownish to bronze or green.

Marsh frog *Rana ridibunda*

The largest frog in Europe, the marsh frog was introduced to Britain from Hungary in 1935 by a zoologist, Percy Smith. He released some in his garden at Stone-in-Oxney on the border of Romney Marsh in Kent. The frogs soon spread to the drainage dykes and ditches that criss-cross the marsh, and in 1937 they were so numerous that their croaking at night brought complaints from local residents. Although the frogs have spread beyond the marsh to adjoining areas of Kent and Sussex, attempts to introduce them to other parts of Britain have not been very successful.

Marsh frogs like to bask in the sun, but quickly take to the water if disturbed. They feed mostly on land, mainly eating flying insects. At breeding time in late May or early June, the males gather in the water in small groups and croak loudly, fully distending their vocal sacs, as they call to attract females. The spawn is deposited in small clumps among water-weeds, and the tadpoles are rarely seen. Young frogs emerge from the water in September and October. Romney Marsh has long been used for sheep grazing, but a change to arable farming is altering the dyke system and could affect the marsh frog's survival.

Food is caught on the frog's sticky tongue. It feeds on insects such as mayflies and other small creatures such as worms.

Found in drainage dykes and ditches only in Romney Marsh and the surrounding area.

At night a toad's eyes have large, circular pupils. In daylight the pupils contract to slits bordered by a golden iris. Toads walk, they do not leap as frogs do, and are slow and clumsy in comparison.

Most predators, weasels for example, leave toads alone because when threatened they thrust poison glands on the back towards the enemy. The glands secrete a weak, distasteful and odorous poison. Some grass snakes eat toads.

Every spring toads migrate to regularly used breeding ponds for spawning. Males, which emit sharp croaks, clasp females tightly from behind and fertilise spawn as it is laid in long strings that wrap round water plants.

In winter, from about mid-October to mid-March, toads hibernate under logs or stones, singly or in large groups. They usually find dry places not far from their breeding ponds.

Warty skin

Long sticky tongue

Common and widespread, but not in Ireland. Frequent in gardens, often in dry places.

The common toad has a dry, warty skin and is generally brown, often with darker spots, but colouring may be yellowish, greyish, reddish or olive-green. A toad feeds on living prey, such as worms, which it seizes with its long, sticky tongue. Female up to 4 in. (10 cm) head and body. Male smaller.

Toads are becoming increasingly common in parks and gardens that have ponds where they can breed. Many natural ponds are now polluted or filled in.

Common toad *Bufo bufo*

Toad migration in March or early April can be a spectacular affair. Within the space of a few days, hundreds of common toads leave their hibernation places and make for their breeding pond, on their way climbing walls and other obstacles and crossing roads in a mass. Many are killed, especially by vehicles, and in some places warning signs are erected on busy roads.

Common toads like to spawn in fairly deep water. Strings of spawn (eggs) are usually about 7–10 ft (2–3 m) long. Eggs develop into tadpoles and then young toads in about 10–16 weeks depending on the weather. Young toads leave the water in June or July, only 1 in 20 surviving to become adult. Males are ready to breed at about three years old and females a year later, so males heavily outnumber females at breeding time.

When spawning is over, adults leave the ponds to live alone through the summer. Most of the day is spent under logs, but the toads emerge in the evening to catch insects and other small animals. They sit and wait for prey to come within range of their long tongue, which is rooted at the front of the mouth and can be extended for nearly 1 in. (25 mm). In captivity toads may live for 20 years, but in the wild not usually more than ten.

207

Because its hind legs are shorter than those of the common toad, the natterjack can crawl faster. It is often active in daylight.

The spawn is laid in strings about 3–7 ft (1–2 m) long in shallow water, wrapped round plants. The strands have single rows of black eggs, not double rows like common toad spawn.

Midwife toad
Alytes obstetricans

For a few weeks in late spring and early summer, a male midwife toad carries eggs from the female twined round its back and hind legs. It takes them to water when they are ready to hatch. One small colony is known in Britain, accidentally introduced in the 1800s with water plants.

A burrowing natterjack toad digs with its forelimbs, throwing soil behind in the same way as a dog.

When frightened, the natterjack, like the common toad, adopts a defensive posture. It arches its back towards an attacker so that the poison-secreting glands on its back are uppermost.

A yellow line down the middle of its back distinguishes the natterjack toad from the common toad. It is also slightly smaller and has a shinier, smoother skin. Adults vary in colour from yellowish-green to olive-green. Head and body 2½ in. (64 mm).

Yellow stripe

A natterjack toad is fully grown when four or five years old. *Calamita*, the species name, is from *calamus*, a reed. Natterjack toads often hide among reeds.

Found on sandy heaths and coastal dunes. Common in a few restricted localities.

At breeding time, from March to midsummer, males call loudly to attract females. Each distends its single throat sac to give a penetrating, rattling croak.

Natterjack toad *Bufo calamita*

The name of the natterjack toad is probably derived from the Anglo-Saxon word *naeddre*, which meant a serpent or crawling creature. The addition of 'jack' may refer to the toad's small size, as it does in jack-snipe, for instance. The natterjack is found only in sandy places, mainly in coastal dunes in East Anglia and north-west England, but was once common on southern heathlands. Now its numbers are declining and it is protected by law.

Natterjack toads dig burrows in soft sand, and often shelter there in a group. They generally emerge to forage at night, feeding on insects and other small animals. Winter is spent buried 1–2 ft (30–60 cm) deep in the sand. Spawning begins in April and continues until June or July. During this time the males sit in the water round the edges of the shallow, sandy pools used for breeding and croak loudly. Their night chorus can sometimes be heard more than half a mile (1 km) away. Many tadpoles die because their pond dries up before they become toadlets. Most toadlets leave the water in June or July.

Although natterjack toads have poisonous skins, some predators – seagulls and crows, for example – have learned to eat them leaving the skin untouched. Toads may live 10–12 years.

209

Newts usually hibernate on land during winter. They often choose a damp cellar or a garden corner, but are never far from water.

Like all newts, common newts are greedy feeders. With their sticky tongues they catch slugs, worms, insects and even other newts. They swallow their prey whole – including snails in their shells.

Great crested newt
Triturus cristatus

The largest British newt, with a slimy, warty skin, blackish above with a black-spotted golden-yellow belly. Males have a high, toothed crest and a silver-streaked tail. The species is declining and is protected by law. About 6¼ in. (16 cm) long including tail.

Palmate newt
Triturus helveticus

The smallest British newt, olive-brown above with a dark streak across the eye. Breeding males have webbed hind feet, a low, smooth crest and a short filament on the tail. About 3 in. (75 mm) long including tail.

The tiny young leave the water at the end of the summer, after they have changed from tadpoles into miniature newts.

At breeding time male common newts have a bright orange underside and spots on the throat as well as the belly. Male palmate newts have a whitish-yellow underside and plain pink throat. The females of the two species are drabber and hard to tell apart.

Smooth, soft skin

Female

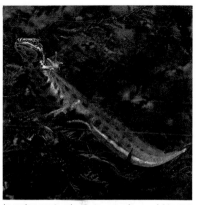

In spring newts migrate to water for courtship and spawning. Males develop brighter, glossier breeding colours and a high crest along the back and tail.

The common, or smooth, newt is the most widespread British newt. Like all newts it has a smooth, soft skin and a tail flattened at the sides. On land it is yellow-olive in colour, with a spotted belly. Females are duller than males. About 4 in. (10 cm) long from head to tail tip.

Common newt *Triturus vulgaris*

Although common newts live in their breeding pools for most of the spring, they spend summer and autumn on land. Most newts also hibernate on land during winter. Open woodland and scrub or lush pasture, with suitable breeding ponds near by, are typical places to find newts, and they are becoming common in gardens. By day they hide under stones or logs or in thick grass. On damp nights they emerge to hunt for slugs, worms and insects, tracking prey by both sight and scent. Newts look something like lizards but have no scales and move very slowly. They never bask in the sun as lizards do.

Early in spring newts move to the water to breed. The eggs are individually wrapped in water plant leaves. Adults return to land in early summer, leaving the tadpoles to develop into fully formed but tiny newts which emerge to live on land at the end of summer. They stay there until two years old, when they return to the water to breed. Common newts are the most abundant, but palmate newts tend to take over in heathland or uplands. Great crested newts are found in lowland England but are becoming rare. If it escapes predators such as hedgehogs, rats and grass snakes, a newt can live about ten years.

Common and widespread, mainly in lowland Britain. The only newt found in Ireland.

A male common newt has a low ridge of skin running along its back in summer, the remains of its spring breeding crest.

211

A rockery or a rough brick wall provides crevices for frogs, newts and toads to hide in by day. Where there is plenty of undergrowth, small mammals such as bank voles and wood mice may also move in.

In a shallow corner the water gets warmer and speeds tadpole development. If there are goldfish in the pond, they will eat large numbers of tadpoles.

Frogs like shallow water, about 4 in. (10 cm) deep, for spawning in February or March. They spend the rest of the year in damp places near the water.

Birds often drink or bathe in shallow garden ponds. Blackbirds will also feed on tadpoles.

A garden pond as a nature reserve

Garden ponds have become life-savers for Britain's frogs, toads and newts in the past 50 years, since so many farm and village ponds have been filled in or become polluted. Garden ponds also benefit the gardener because the amphibians they attract prey on creatures such as slugs, snails and insects. Ponds will attract bathing birds and animals such as foxes, which drink from them. Small mammals such as hedgehogs often fall in and cannot climb out if the pond has smooth sides. Wire mesh hanging in the water in one corner enables them to escape.

Whether it is made from an old sink or shaped plastic, a pond soon becomes colonised by insects such as pond skaters. A few jars of natural pond water tipped in will add plankton, pond snails and other creatures that are food for larger creatures such as water beetles. Frogs and toads may arrive of their own accord, but a colony can be started by introducing spawn in spring and leaving it to develop. If all goes well, adults will return to breed two or three years later. Toads generally prefer deeper ponds than frogs. Newts may find their own way, or can be caught as adults and released in the pond in spring. Great crested newts are legally protected, and must not be captured without a licence from English Nature.

Newts spend most of the year on land in damp places – under logs, for example. They hunt for prey such as slugs at night.

Long, damp grass near the pond provides shelter for young frogs, which leave the water in June or July.

213

Females usually have up to 15 young at a time, but up to 20 is possible. Although young adders sometimes hide under their mother, she gives them no parental care and they disperse not long after birth.

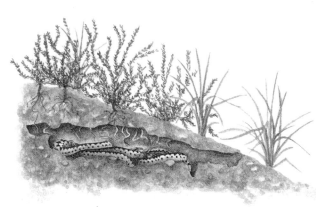

From October until March adders hibernate in a cool, dry place. They use an existing crevice or hole in the ground. A small group will often share a good place.

Very dark adders are not uncommon. Totally black adders, with no zigzag marking distinguishable, are found in some areas.

Adders usually hunt by day. They kill with a bite from hollow, hinged fangs, which inject poison from venom glands in the upper jaw.

If the victim does not die at once, the adder trails it, using its forked tongue to follow the scent. The snake has heat sensors on its snout with which to detect the body.

The snake folds back its fangs before swallowing. It draws in its prey by working its loosely hinged, wide-opening jaws. Each half of both the upper and lower jaws can be moved independently.

Females are usually duller and browner than males, with less contrast between markings and background colouring. They are also fatter, and may be up to 30 in. (76 cm) long.

V marking

The best time to see adders is in spring, when they often bask in groups. Basking adders prefer gentle warmth, not the strong heat of the midday sun.

Zigzag line

Widespread, but distribution patchy. Found in hedgerows, farmland, open moors, woods.

In spring male adders may be seen 'dancing' (wrestling to win a female). They rear, sway, writhe and race over vegetation as each tries to force the other to the ground. Adders vary in colour and markings, but a dark zigzag line along the back and a V mark on the back of the head are characteristic. Male about 24 in. (60 cm) long. Female often longer.

Adder *Vipera berus*

According to an old legend, a female adder swallows her young when danger threatens. What actually happens is that the young snakes hide under their mother's belly. The adder, or viper, is Britain's most widely distributed snake and the only one that is poisonous. But it is very timid and normally flees from humans before anyone gets close enough to provoke it into biting. The bite is rarely fatal but needs hospital treatment as soon as possible.

Adders like open places such as heaths, moors and scrub-covered hillsides, and are sometimes found among sand-dunes. They shed their skins from time to time, and cast skins may be a sign of their presence. On farmland they are the farmer's friend, eating small mammals such as mice and voles. They also eat lizards, frogs and toads if nothing else is available, but do not eat every day; a large meal may last a week or more. Male adders emerge from hibernation in February or March, females a little later, and courtship reaches its peak in April. Females usually become pregnant only once every two years. The young are born fully formed in August or September and take two or three years to reach maturity. Adders have few enemies apart from man, and may live for nine or ten years.

Lizards are the smooth snake's chief prey. It will also eat small mammals, probably taking them in their burrows. Young snakes eat insects and spiders.

The snake is not poisonous, nor is it a constrictor. It holds its prey in its coiled body while it gets into position to swallow the victim head first.

The smooth snake can escape heath fires by hiding among grass roots or underground. But it may die of exposure or starvation in the burned, barren landscape that results.

Up to 15 young are born at a time, each within a thin membrane that ruptures at once. They are about 6 in. (15 cm) long, with darker heads and spots than adults.

A dark patch on the back of the head is shaped something like a coronet, but is not as distinct as an adder's V mark. The eye has a round pupil, not a vertical slit like an adder's.

Like all snakes, the smooth snake is brightest in colour after shedding its skin. Old skin is regularly cast off in one piece, the snake squeezing between roots and twigs to free it.

Eye stripe

Dark spots

The most likely time to see the rare smooth snake is in April or May. It enjoys basking in the mid-morning sun, sometimes on bare ground between heather shoots.

Very rare. Found on some heaths in southern England. Numbers probably declining.

The smooth snake is slimmer than an adder, with smoother scales. It also has a narrower head with a dark side-stripe through each eye. Colouring varies from grey to brown or red-brown, and there are dark spots down the back, sometimes joined as bars. Up to 24 in. (60 cm) long.

Smooth snake *Coronella austriaca*

Lowland heaths of the type found in Surrey, Hampshire and Dorset are the home of the smooth snake, one of Britain's rarest animals. Because it is so rare it is now protected by law, and whether or not it can continue to exist in Britain depends on how much heathland can be preserved. Lowland heaths are under increasing pressure from farmers and foresters, as well as from walkers and picnickers and the ravages of heath fires.

Because the smooth snake is very shy as well as rare, there is still much to be learned about its way of life. It rarely basks in the open, preferring to get the sun's warmth indirectly by lying under a flat stone or similar objects. In spring, when they emerge from winter hibernation, smooth snakes are sometimes seen basking intertwined among heather. Like many other reptiles, individuals have a particular home range and stay in the same area for a long time. Much of their time is spent burrowing underground, and they mostly eat other reptiles such as lizards. They are also cannibals, at least in captivity.

Mating takes place in May. The young are born in August or September and are self-sufficient from birth. Smooth snakes may survive in the wild for 15-20 years or more.

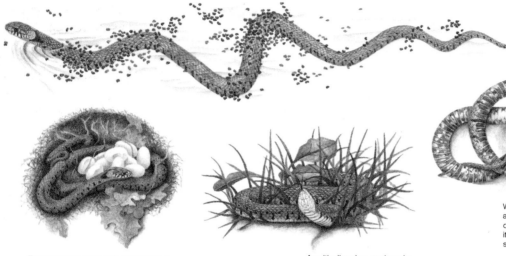

A grass snake swims with sinuous body movements as it hunts in a pond or a stream. It swallows tadpoles under water but takes larger prey ashore to eat.

To lay her eggs, the female burrows into a place where heat is generated, such as a haystack, a manure heap or rotting vegetation. The eggs, up to 40, are matt white and 1¼ in. (30 mm) long.

As with all snakes, each scale on the underside spans the body width. Scales have sharp rear edges and dig into the ground to give the snake a grip as it moves forward.

When threatened by a predator, a grass snake sometimes feigns death by lying on its back with its tongue lolling. It may eject smelly liquid.

The eggs have flexible shells and are held in clusters by a sticky film. Two months after laying, young snakes about 7½ in. (19 cm) long emerge through slits made by an egg-tooth on the snout. The tooth is shed a few hours later.

The pupil of the eye is circular and there is no eyelid. The eye is always covered by a transparent part of the skin. The top of the head is all one colour.

A snake's skin is sloughed (shed) 3–12 times a year. Before it is shed it goes dull and dark and the snake's eyes look misty. Colouring is brightest just after sloughing.

The grass snake's olive or grey-green colouring may be light or dark. The snake constantly darts its forked tongue in and out, especially a courting male.

Black bars

Neck patches

Also known as the ringed snake, the grass snake has two yellow or white patches almost encircling its neck. Black bars almost always mark the flanks, and there are black dots along the greenish back. The belly is patterned with black, grey and white. Female up to 48 in. (120 cm) long, male smaller.

Commonest in lowland areas. Found mainly in damp heaths, woods, or in lush pastures.

Grass snake *Natrix natrix*

Damp grass, ditches, pond banks and slow-moving streams are likely places to see the grass snake, Britain's largest snake. Amphibians such as frogs are its main food, caught on land or in the water. It often feeds in the early morning, and a large meal will satisfy it for a week or ten days. Grass snakes are harmless to humans. Like other reptiles, they spend a lot of time basking in the spring sun after emerging from hibernation in April. Courting and mating follow soon after.

The grass snake is the only British snake to lay its eggs in a place where heat is generated, such as a compost heap. A female will travel more than a mile (up to 2 km) to find a suitable site, and many may be attracted to the same one. Hundreds of eggs may be laid in one manure heap in June or July, giving rise to tales of snake plagues when the young hatch in August or September. Eggs laid in hot places hatch some weeks earlier than those laid in cooler places, and the young have the benefit of the summer sun while growing, giving them plenty of energy to hunt and fatten up for winter. Hibernation begins in October, in wall crevices, under tree roots or in similar places. Snakes not taken by birds, hedgehogs or badgers may live about nine years.

A lizard may be seen with a short tail as it grows a replacement for one that has been shed. Sometimes a new tail grows beside one only partly lost, giving a double tail.

There are usually from five to eight black young – sometimes ten. They are born fully formed, each in a transparent capsule that usually breaks at birth. They may all be born at once, or over a one or two-day period. Common lizards may live for five or six years.

Its wide-spreading, sharp-clawed toes enable a lizard to grip and scale a roughened vertical surface with ease.

Spiders, harvestmen, flies, beetles and caterpillars are among the lizard's prey. It catches prey in its jaws and shakes and stuns it before eating.

The transparent surface layer of skin is shed periodically. It is scraped off in pieces, making the lizard look ragged until moulting is completed.

Dark back stripe

Male

The common lizard is agile and sharp-clawed and can easily climb a tree. It merges well with its background until its rapid, darting movements give it away.

Female

Dull brown is the typical lizard colouring, but it may be tinged red, yellow, grey or green. There is almost always a dark back stripe, and often dark side stripes with white edges. Stripes are sometimes broken. The female is fatter than the male and usually paler. About 6 in. (15 cm) long.

Common lizard *Lacerta vivipara*

Rustles in the undergrowth may be the first sign of a common lizard's presence as it scampers up a bank to hide. It is very nimble and disappears rapidly when disturbed. Lizards live in heathland, sand–dunes, grass or scrub–covered banks and on high moors, and where conditions are ideal large colonies may be found. They emerge from hibernation early in spring, and at first bask in the sun a good deal, but as the weather gets warmer they need to bask less. They court and mate in April and the young are born in midsummer. In the cool of September and October, lizards need to bask more often. Before autumn ends they retire into cracks and under stones to hibernate.

Common lizards eat a variety of small creatures, spiders particularly, which they hunt throughout warm days. At night and on cool days they remain hidden. Lizards themselves fall prey to many animals, including smooth snakes, adders, rats and birds – particularly kestrels. If a predator seizes a lizard's tail, it can shed it and so escape; the tail may even be shed if the lizard is merely threatened, to divert attention while it scuttles to safety. A new tail slowly grows from the stump, but is never as long or as perfect as the original.

Widespread in open places throughout Britain. The only lizard in Scotland, Ireland.

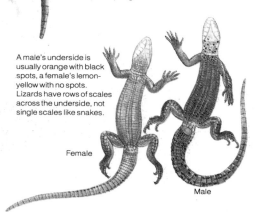

A male's underside is usually orange with black spots, a female's lemon-yellow with no spots. Lizards have rows of scales across the underside, not single scales like snakes.

Female

Male

Adult males are aggressive in the April-June breeding season, threatening each other and picking fights. One may display its conspicuous flank colouring to intimidate an opponent.

A female shows aggression by opening her mouth and raising and shaking her forefeet alternately. She may do this to warn off an unwanted male.

Before mating, the male chases the female and grips her first by the tail, then further forward and finally by the middle of the back.

Sand lizards that live among dunes differ from heath lizards in having unbroken stripes and fewer spots. The male's flanks are lime-green.

The oval eggs, about $\frac{1}{2}$ in. (13 mm) long, hatch after two or three months. The newly hatched lizards are $2\frac{1}{2}$ in. (64 mm) long.

In late May or early June the female digs a hole in a sunny place in soft sand, lays the eggs there and covers them up.

The vivid green flanks of the male sand lizard are breeding colours developed in late spring. The green fades gradually as the summer passes.

Light stripes

Flank patches

Female

Male

More heavily built and solidly marked than the common lizard, the sand lizard has two light stripes, sometimes broken, along the back, with dark spots between them. Black patches on the flanks have pale centres. Males are greenish, females brownish. About 7 in. (18 cm) long including tail.

In the north-west of England prefers coastal dunes, but in the south sandy heaths.

Young sand lizards can be distinguished from common lizards by their more conspicuous spots.

Sand lizard *Lacerta agilis*

Dry, open country is the home of the sand lizard, chiefly the sandy heaths of Surrey, Hampshire and Dorset, although a small population lives on the coastal dunes near Liverpool. As the heaths give way to farming and forestry and the sand-dunes become less secluded, sand lizards have become scarce and are declining still. Since 1975 they have been protected by law.

The sand lizard emerges from its hibernating burrow, dug deep in the sand, in March or April, and spends some time basking and shedding its old skin. Males normally emerge first, their dull winter colours soon being replaced by emerald-green flanks. In April and May the males fight fiercely for dominance, the victors pairing up with females. Adult lizards feed well during summer to build up fat reserves for the coming winter. Large insects such as beetles and grasshoppers are the main food, and they are often dismembered before being eaten. Sand lizards may start hibernating as early as the end of August. Young lizards are hatched in late August or September and grow rapidly before they hibernate in October. Sand lizards need at least two years to reach maturity. They may live to be seven or eight if they escape the snakes, rats and birds that prey on them.

223

If it is seized, a slow-worm can shed its tail and escape. The stump eventually heals and the tail grows again, but never to its original length.

Slugs are the slow-worm's favourite food. It seizes one by the middle and then swallows it whole. Insects, spiders and other small creatures are also eaten. The slow-worm takes only live prey, not carrion.

Slow-worms like to bask in partial sunlight, not in the open. They sometimes drape themselves among the twigs of old heather, or lie under corrugated iron or thin stone slabs.

Much of the slow-worm's time is spent underground. It burrows in soft soil, such as a garden compost heap. Although it is frequently found in an ants' nest, it does not seem to eat ants very often.

Each young slow-worm is born in a membranous egg that it breaks open within seconds. It is golden-yellow above and black below, and 3 in. (75 mm) long.

Sleek skin

Many older male slow-worms have bluish spots on their backs and flanks, especially near the head. No reason for this colour variation is known.

Female

Dark stripe

Male

Slow-worms are sleek, shiny, snake-like lizards. Adults vary in colour from greyish to light, dark or coppery brown. Females have dark brown flanks and belly with a paler back, sometimes with a dark stripe. Males are more uniform in colour. Up to 18 in. (45 cm) long; just over half is tail.

Widespread and common but shy and not often seen. Rarely basks in the open.

Like most lizards, the slow-worm has eyelids and can close its eyes, and also has a broad, flat tongue. But it differs in having no visible earholes.

Slow-worm *Anguis fragilis*

At one time any creeping, serpent-like animal was called a worm, which is how the slow-worm – or blind-worm – got its name. It is in fact a legless lizard but is often mistaken for a snake and consequently killed, although it is harmless. As slug-eaters, slow-worms are an asset to any garden. They usually move slowly and deliberately, but can move fast if disturbed.

Slow-worms live on sunny banks and hillsides where there is good cover such as grass, scrub or stones. Their shed skins, left in fragments, may be found occasionally. Hibernation underground begins in October and ends in March, when slow-worms emerge to bask in the early spring sun. Mating takes place in April and May. Males often fight at this time, seizing each other on or near the head and sometimes inflicting serious bites. From 6 to 12 young are born in August or September or, more rarely, in the following spring.

Young slow-worms have many predators, including frogs and toads. They take about three years to mature. Slow-worms can live longer than any other lizards, and one in captivity has reached its fifties. Such an age is unlikely in the wild, where enemies include hedgehogs, adders, rats and kestrels.

225

A common lizard keeps its body heat at about 86°F (30°C) for as long as it can by basking in the sun and cooling in the shade. In weak sun it flattens its body to expose as much to the sun as possible. On hot days it turns its face to the sun, so offering a smaller surface to receive warmth. Basking exposes the lizard to kestrels and crows; it must always be ready to dart into a crevice.

Pieces of corrugated iron and other metal, such as farm machinery and garden tools, become very warm in the sun and reptiles can bask under them secure from attack.

Slow-worms bask in semi-shaded places such as in long grass or the shadow of a wall. They absorb less warmth there but are less visible to predators.

A crevice in a drystone wall or a burrow at its base offers a common lizard a place to hide at night and on sunless days, as well as somewhere to hibernate from October to March.

Why snakes and lizards bask in the sun

Although reptiles such as snakes and lizards are usually described as cold-blooded, they spend much of their lives with their bodies nearly as warm as in a warm-blooded animal. What they lack is a means of producing and maintaining their own body heat. To reach a body temperature at which muscles, senses and digestion are fully active (about 77–90°F, 25–32°C), they have to rely on outside heat – either conducted from their surroundings or, more often, direct from the sun's rays.

In Britain, reptiles are dormant in winter because they cannot reach their working temperatures. In spring they emerge on sunny days, but the sunshine is so weak they need to bask almost all the time. As the sun gets stronger, basking time is shorter – mainly early and late in the day – and there is time for hunting and breeding. By midsummer basking is hardly necessary, but as autumn comes the need increases, and finally the animals are forced into hibernation. Reptile species that bear live young can usually survive in cooler climates than those that lay eggs; by basking during pregnancy, they can get enough heat to ensure faster development of the young but warmth cannot be ensured for eggs. This is one reason why the adder and common lizard, which bear live young, can live farther north than the grass snake and sand lizard, which lay eggs.

Heaps of stable manure or rotting plants generate heat, which makes them suitable basking spots for grass snakes. The snakes often lay their eggs inside a heap, where the warmth speeds incubation.

Tracks and signs in downland and pasture

Most wild animals stay under cover during daylight hours, and venture out only after dark. As a result, few are likely to be seen in open fields and downland during the day. To the observant eye, however, there are usually plenty of signs that reveal the presence of quite a number of species. The signs described here are typical of the type of country shown, but are not the only ones that may be found there.

Because fields and grassland are frequently grazed by farm animals, the footprints, droppings and hairs of wild animals can easily be confused with those of domestic species. Sheep and deer, for instance, both have cloven hoofs, and it is very difficult to distinguish between their tracks. Generally, however, deer tracks in a particular spot are fewer in number than those of sheep, as sheep live in large groups. Deer are surprisingly common, even in heavily inhabited countryside, although they usually stay hidden from sight. Red, sika, fallow and roe deer are the species whose tracks are most likely to be confused with those of sheep. Pastureland where sheep graze is usually very close-cropped and unlikely to support many small mammals, except round field edges where hedgerows and longer grass afford food and shelter.

Rejected ragwort
A field with a lot of ragwort has probably been grazed by horses. Both horses and cows find ragwort distasteful but a horse can be more selective in its eating. It bites off plants between its teeth, whereas a cow pulls them up in clumps by curling its tongue round them.

A fox's leavings
The hind-leg bones and the foot of a rabbit are likely to be the remains of a fox's meal, and often have a distinctive, musky fox scent. They may have been dropped by a fox or, if the bones are picked clean, by a scavenging crow.

A fox's droppings
Fragments of bone, matted fur and seeds are contained in the droppings of a fox; the droppings have a very characteristic twisted tail.

Trampled trails in the grass
Beaten tracks in the grass that look like footpaths may be animal trails. If they pass under fences or low bushes and shrubs, they have been made by an animal. Footpaths made by humans go over or round such obstacles, or trample them down.

Sheep trails are usually narrow. They often cut deep into soft ground and expose the bare earth.

Badger paths are usually well-worn tracks about 6 in. (15 cm) wide. Normally the grass is flattened, but no earth is exposed.

How to recognise hairs

Where animal trails pass under a chain-link or barbed-wire fence, hair can be scraped off an animal's back. Hairs caught on the top strand of a fence are likely to be from horses or cows, which rub their necks along it, or from deer that have leapt over it. Small deer also squeeze under fences.

Cow hairs are fairly short and soft, and mat together like felt.

Deer hairs are stiff, straight and bristly.

Top wire

Horses often catch long mane and tail hairs on barbed wire.

Badger hairs are wiry and 2–3 in. (50–75 mm) long. They are white, with a black zone near the tapered end.

Rabbit hairs are fine, soft and fluffy. They are about $\frac{1}{2}$ in. (13 mm) long, and grey with a fawn tip.

Bottom wire

Fox hairs are about 1 in. (25 mm) long. They are straight and red-brown or grey-brown with a pale tip.

Where to look for footprints

A small patch of mud, often found near a gate or on a rutted path, is a good place to look for animal tracks. The cloven hoof-prints of sheep and deer are difficult to tell apart, but there may be other identifying signs near by, such as wool or hair.

Hind

Fore

A badger's fore and hind feet can be mistaken for the tracks of two different animals.

A dog's pawprints are broad and often have the front toes splayed.

A fox's tracks are narrow, with the front toes close together.

Sheep tracks have one half of the hoof larger than the other.

Deer tracks have both halves of the hoof the same size but often more widely separated.

Tell-tale wisps

Where sheep have grazed, pieces of greyish-white fleece often get caught on brambles and thistles.

Honey for the badger

Common wasps and some wild bees make nests underground in holes or burrows, often in a bank. A dug-up bees' nest is the work of a badger, which eats the larvae and stored honey. Its shaggy coat protects it from stings.

Bottles as deathtraps

Small mammals will go into bottles to look for food or simply to explore. They are often unable to get out again, so die of starvation and cold. Thousands are trapped in this way every year, and up to 28 have been found in one bottle. Discarded bottles with animal remains are most likely to be found in lay-bys. Where only the skeletons are left, the skulls can be used to identify the trapped animals.

A shrew skull (above and right) is narrow and pointed and does not have prominent cheek bones. The teeth are tipped with red and are continuous along the jaw.

A rodent skull (above and right) such as that of a vole or mouse has prominent cheek bones and eye sockets, and there is a gap between the incisor teeth at the front and the molar teeth at the back.

A mouse (or rat) jaw has knobbly molar teeth; if the teeth are removed, each leaves several tiny root holes.

A vole jaw has a zigzag pattern on the grinding surface of its molars; if the teeth are removed, each leaves one big, ragged hole.

Wayside warehouses

An old bullfinch or blackbird nest containing chewed hawthorn berries or rose hips may have been used as a feeding place by a wood mouse. Small mammals such as mice and voles sometimes climb into bushes to feed on fruit.

A cardboard canopy

Large pieces of rubbish – corrugated iron sheets or cardboard boxes, for example – provide shelter for field voles, whose well-worn runway systems can often be found underneath. There may also be a nest made from finely chewed plants.

Animal traps and traces in roadside lay-bys

Lay-bys are often very untidy places, and the rubbish that people leave there is all too evident in winter when the blanket of green vegetation dies back. To the local mammals, the rubbish can be a blessing or a disaster. Picnic remains and other food may be welcome to a hungry animal, particularly in frosty weather when natural food is hard to find. Unsightly junk may also provide a cosy and weatherproof nest site.

Discarded bottles and drink cans, however, are often death traps for the small mammals that venture into them. More than 50 mice, voles and shrews may die in bottles in a single lay-by in the space of a few months, and up to five different species may be trapped in one milk bottle. Their corpses provide a feast for flies and carrion beetles but are a health hazard for humans, as well as a sad reminder of the suffering caused by people too thoughtless to take their rubbish home. Larger animals are also at risk from litter. A hedgehog can get its snout stuck in opened bean or soup cans and plastic cartons. Deer and sheep step on rusty tins and pick up an uncomfortable anklet that may cause septic wounds. Many animals, including dogs, risk dangerous cuts to feet and noses from broken glass and jagged metal.

Homes in hedgerows and banks

Litter in lay-bys is a source of extra food for mammals in winter. Where there is also a hedgerow with plenty of cover, several kinds of burrowing mammals may live in a lay-by.

A rabbit hole (above) is normally about 3 in. (75 mm) across. There are usually several near each other, with a lot of bare earth around. There are also small, dry, spherical droppings.

The hole of a brown rat (left) is about 1½ in. (40 mm) across, usually with a narrow, beaten trail leading to it. Sometimes there are a few oval droppings near by.

A badger hole (left) is about 12 in. (30 cm) wide and is generally in a slope on light, well-drained soil. Usually, the hole does not smell and has no food remains outside; there may be loose soil containing badger hairs, thrown out with discarded nesting material.

A fox hole (right) is about 10 in. (25 cm) across, with a long mound of excavated soil – and sometimes food remains such as bones – outside. Often there is a strong smell of fox in the vicinity of the hole.

Vigour from decay

The rotted remains of litter add nitrogen to the soil. The result is flourishing clumps of elder bushes and nettles, which like rich soil.

Nettles

Elder flower and berries

Cans that kill

Like bottles, drink cans are traps for small animals. But because the entrance hole is so small, they usually catch only shrews, which are very tiny with a narrow head.

Tracks and signs by the waterside

The muddy bank of a river or stream shows up animal footprints particularly well. So does the soft, moist soil of a water meadow – a riverside that in winter is deliberately flooded to keep the grass green and provide early grazing. The softer the mud, the finer the detail of the prints and the better the chance of very small prints showing up.

Animal footprints seen in the field are not always perfect specimens. Sometimes only the claws or toes may mark the soil, especially on dry ground, and there may be no print from the palm or sole of the foot. The fore feet of water voles and rats, for instance, often leave only sets of four small toe marks in an arc. Where several animals have crossed the ground, or one has moved repeatedly over the same area, there is likely to be a mass of confusing, overlapping prints. But there may be signs near by to give a clue to the animal's identity: food debris, for example, or a pile of droppings. As well as animals that live beside the water, look for the tracks of animals that come to drink at the water's edge, such as badgers, foxes and deer. Bird footprints, recognisable by the very long middle toe, are often numerous, particularly those of ducks, herons and moorhens.

Footprints in focus

The soft, slimy mud at the water's edge shows very fine detail of even quite small footprints such as those of water voles, the ones most likely to be seen. They can be confused with the footprints of the brown rat, also seen near water. Mole footprints are occasionally seen in wet meadows, after flood water has forced them to leave their burrows.

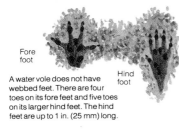

Fore foot

Hind foot

A water vole does not have webbed feet. There are four toes on its fore feet and five toes on its larger hind feet. The hind feet are up to 1 in. (25 mm) long.

Untidy voles

Chewed plant fragments, bitten-off stems and shreds of pith signify the presence of water voles, which feed on sedges and other waterside plants. Look out for tracks, droppings and also burrow entrances near by.

Cows may graze on waterside plants, but do not usually leave fragments lying about. Their cloven hoof prints, up to 3 in. (75 mm) long, are evident.

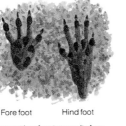

Fore foot   Hind foot

A brown rat has four toes on its fore feet and five toes on its larger hind feet, but its prints are slightly larger than a water vole's. The rat's hind feet are 1⅓ in. (35 mm) long, or more.

Because a mole tends to walk on the inner side of its fore feet, its fore prints show only the marks of its five claw tips, not the rest of its foot. As its legs are short, its belly makes a drag mark in the mud.

Water vole tracks are usually in groups, overlapping each other, around plant fragments. There may be wet, shiny patches where water has run off the fur.

Water vole droppings are oval and about ½ in. (13 mm) long. They are dark green or brown, usually in groups of about half a dozen, with vole prints all round them.

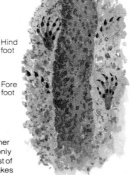

Hind foot

Fore foot

Geese among the mammals

Canada goose droppings – large, soft and green and white – are often seen in heaps beside water. Mammal droppings differ in having no white in them, nor are they as long and thin as the goose droppings.

Burrows at the water's edge

Burrows may be found where there are low banks of soft earth by slow-flowing rivers. Water vole burrows and perhaps a kingfisher's nest-hole are the commonest.

A kingfisher's nest-hole is similar in size to a water vole burrow but is at least 24 in. (60 cm) above water level, and normally in a bare, vertical bank.

The entrance to a water vole burrow is a hole about 2 in. (50 mm) across. It is in a waterside bank at or near water level.

Territory markers

The droppings of aquatic mammals such as mink and otter may be found on a fallen tree or boulder at the water's edge. The animals leave the droppings as scent markers to define territory. Tracks in the mud near by help identification.

Mink droppings are smaller than otter spraints, but may also contain scales and bones. They are foul-smelling and when fresh are dark green.

Otter droppings (spraints) are dark and slimy with a strong, somewhat fishy smell. Fish scales and bone fragments may be visible.

Fore foot

Hind foot

Otter tracks show five-toed webbed feet. They are up to 2⅜ in. (60 mm) in length and about as wide as they are long. Drag marks made by the animal's tail can often be seen between the footprints.

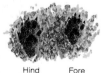

Hind   Fore

Mink tracks show five-toed feet up to 1½ in. (40 mm) long. Web marks are not always visible. On the fore feet one toe is set back and close to the pad.

233

An ever-ready larder

In mixed woods, squirrels often choose to live in conifers such as Scots pines, which provide them with year-round food and shelter. Look for chewed cones scattered beneath pines or other conifers.

A squirrel (red or grey) bites the scales off a ripe cone to get at the seeds inside. Chewed cores and neatly bitten-off scales are dropped.

Crossbills also drop discarded pine or spruce cones below trees. Usually the scales are not torn off but prised open with the beak and split.

Tracks in the mud

A muddy wheel rut in a woodland ride is a good place to look for animal tracks, particularly those of deer. Look, too, at the edge of a ditch where animals jump across.

Saplings spoiled by deer

In April roe bucks fray the bark of saplings such as birches when they rub off velvet from their new antlers. The ground below is churned up by their hooves. They also rub scent on saplings from April into July.

Red deer hoofprints are roughly 3¼ in. (83 mm) long and 2½ in. (64 mm) across.

Fallow deer hoofprints are roughly 2½ in. (64 mm) long and about 1½ in. (40 mm) across.

Hedgehog sign

Hedgehog droppings are sometimes seen on pathways. They are usually single, black, crinkly, about 1½ in. (40 mm) long and often studded with the remnants of chewed beetles.

Roe deer hoofprints are roughly 1¾ in. (45 mm) long and 1⅓ in. (35 mm) across.

Muntjac hoofprints are less than 1¼ in. (30 mm) long and 1 in. (25 mm) across.

Dew claws

When deer jump and land in soft mud, their hoofprints are widely splayed and the dew claws are often imprinted.

Tracks and signs in mixed woodland

Mixed woodland often provides shelter for more different kinds of animals than any other type of country, but it takes patience and a keen eye to find their signs. Tracks and trails do not show up well in leaf litter, and in autumn are soon hidden by falling leaves. But there are often animal footprints in the muddy rides and bare patches of soil, although it is rarely possible to follow a trail far before it becomes lost in the undergrowth. There are also animal signs such as hairs, droppings and food remains to be seen, although they are not easy to pick out amid the entanglement of trees and bushes.

On dry woodland soils, paw tracks – especially dog tracks – do not show the whole print, only the two front toes, and could be mistaken for the cloven hoofprints of deer. Usually the distinct claw marks to be seen at the tips of the toes of pawprints help to avoid confusion. The droppings of different species have distinctive shapes, but the colour and texture depend very much on what the animal has been eating, which often varies according to the time of year. Generally, groups of smooth, uniform pellets are those of plant eaters, and ragged pellets found singly or two or three together are those of carnivores.

Thickets that shelter deer

Deer often shelter beside a windproof holly thicket or dense coppice, or under yew trees. Piles of droppings (fewmets) accumulate there.

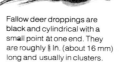

Fallow deer droppings are black and cylindrical with a small point at one end. They are roughly $\frac{5}{8}$ in. (about 16 mm) long and usually in clusters.

Muntjac droppings are black and less than $\frac{1}{2}$ in. (about 10 mm) long. They are round or slightly elongated and usually in clusters.

Red deer droppings are black and cylindrical with a small point at one end. They are up to 1 in. (25 mm) long and usually in clusters.

Roe deer droppings are oval and roughly $\frac{1}{2}$ in. (10–15 mm) long. They are usually in clusters and may be black or brown.

Hazel tree harvest

Hazel nuts from woodland and hedgerow trees are a major food for rodents, which gnaw them in different ways depending on the species. Study the discarded remains of nuts to identify the eater. You may need a lens to see the toothmarks.

A bank vole gnaws a hole with a regular, clean-cut edge. It leaves few if any toothmarks on the nut surface.

A squirrel's jaws can shatter a nut in the same way as nutcrackers can. It leaves fragments of irregular shape with jagged edges and no obvious toothmarks.

A wood mouse gnaws a neat hole with an irregular, chamfered edge. There are traces of toothmarks on the nut's shiny surface.

A dormouse gnaws a hole in the side of the nut and then enlarges it by turning the nut round and scooping the edge with its teeth. This leaves toothmarks on the cut edge but few on the nut surface.

Nests in coppice stools

Trees are coppiced by being cut off near the ground. This encourages the growth of clusters of slim, straight shoots for use as poles. Stumps (stools) accumulate a mass of dry, dead leaves that are ideal places for dormice to hibernate in winter. Several other species, such as wood mice, may nest there too.

Old bones for new

Antlers lying on the ground after being cast by deer are often gnawed by rodents, as they are a source of calcium, which helps to strengthen bones.

Reading the stories in the snow

A layer of snow will show up the tracks of all kinds of animals not normally seen and whose presence was unsuspected. In snow, no animal can move anywhere without leaving its footprints, so snow provides an excellent opportunity to gain information about the movements and habits of wild creatures. Light snow, or even heavy frost, shows up individual tracks quite well. Thicker snow that blankets the ground often allows trails to be followed for considerable distances. In really deep snow, however, few trails are likely; it is so difficult to walk in that not many animals venture out.

Generally, only the tracks of fairly big animals – squirrels and larger creatures – are found in snow. Smaller animals such as mice and voles stay underneath the snow blanket because their food is there, and it is also warmer and more protected from icy winds. Trails in the snow often lead to where an animal has fed – where a rabbit has scraped through to find grass, for example. With the trail of a predator there may also be tracks of the prey, as well as its remains. Other signs of an animal, such as droppings or stains in the snow where it stopped to urinate, may also be found near its tracks.

Different ways of walking revealed

Trails in the snow usually have a whole succession of footprints, and so give a good opportunity to study the different ways in which animals walk.

A fox trail is almost a single line because a fox puts its hind feet into the prints of its fore feet and draws in its feet below the mid line of its body.

A dog trail has prints staggered to each side of its body line, being relatively wider-bodied and shorter-legged than a fox. It does not put its hind feet exactly into the prints of its fore feet.

A badger walks with its feet pointing inwards; the hind prints often overlap the rear of the fore prints. Often the prints are muddy from soil on the feet. Badgers do not hibernate; trails may be seen between sett entrances.

Hind foot

Hopping Bounding

How speed affects trails

The pattern of an animal's footprints varies according to its gait – hopping, trotting or galloping, for example.

Hind foot (about 3½ in., 90 mm)

Fore foot

A rabbit moving slowly hops along. The prints of its hind feet are just behind the prints of its fore feet. But when it moves fast, it bounds in much longer strides. Its hind footprints appear in front of its fore footprints.

Trotting Galloping

A trotting fallow deer places its hind feet just in front of the prints of its fore feet, producing a staggered trail of partly overlapping prints, all more or less evenly spaced. When galloping, it puts its hind feet well in front of its fore feet. Its prints are in fours, separated by longish gaps. Often prints are splayed and the marks of dew claws show.

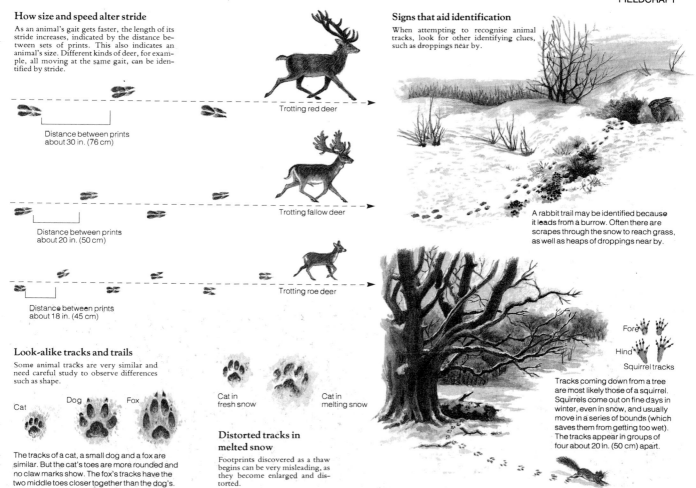

How size and speed alter stride

As an animal's gait gets faster, the length of its stride increases, indicated by the distance between sets of prints. This also indicates an animal's size. Different kinds of deer, for example, all moving at the same gait, can be identified by stride.

Trotting red deer

Distance between prints about 30 in. (76 cm)

Trotting fallow deer

Distance between prints about 20 in. (50 cm)

Trotting roe deer

Distance between prints about 18 in. (45 cm)

Signs that aid identification

When attempting to recognise animal tracks, look for other identifying clues, such as droppings near by.

A rabbit trail may be identified because it leads from a burrow. Often there are scrapes through the snow to reach grass, as well as heaps of droppings near by.

Look-alike tracks and trails

Some animal tracks are very similar and need careful study to observe differences such as shape.

Cat

Dog

Fox

The tracks of a cat, a small dog and a fox are similar. But the cat's toes are more rounded and no claw marks show. The fox's tracks have the two middle toes closer together than the dog's.

Distorted tracks in melted snow

Cat in fresh snow

Cat in melting snow

Footprints discovered as a thaw begins can be very misleading, as they become enlarged and distorted.

Fore

Hind

Squirrel tracks

Tracks coming down from a tree are most likely those of a squirrel. Squirrels come out on fine days in winter, even in snow, and usually move in a series of bounds (which saves them from getting too wet). The tracks appear in groups of four about 20 in. (50 cm) apart.

237

How to identify bones

It is not unusual to find animal bones, or even whole skeletons. Dead animals, especially larger ones, are rarely eaten whole by predators or scavengers because their bones are too big and indigestible. Bones are often carried off and left elsewhere. Even mice will drag small bones away to gnaw them for their calcium content, and the bones of animals that died underground are often dug up and scattered during later burrowing activities.

All land mammals have a similar bone structure, and by studying the size and shape of bones and skulls it is possible to get an idea of the part they played in the animal's make-up, and perhaps to identify the animal they came from. Bones can be a health risk unless well weathered and cleaned of flesh. Handle them with a stick, or protect your hands with a polythene bag or rubber gloves.

THE SKELETON OF A RABBIT

Shoulder blade
A triangular bone with a big ridge across the surface.

Rib
A slender, curved bone with no smooth end for forming a joint.

Vertebrae (backbone sections)
Each bone has a large central hole for the spinal cord to pass through, with an upright prong above it; there are also side prongs. The prongs are more prominent on body sections than on neck sections.

Skull

Juvenile bones
In young animals, the ends of long bones are not firmly joined to their shafts. The cap is often detached, leaving a rough end.

Neck section

Body section

Pelvic bones
The two pelvic bones, one on each side, are loosely joined together. Each has a large hole in it and also a socket for the thigh bone.

THE HIND LEG OF A DEER

Ankle

Chewed bones
The ends of old bones often show toothmarks because rodents gnaw them to obtain calcium. A sawn or chopped bone is either a cow or sheep bone from meat sold by a butcher.

Upper forelimb bone (humerus)
A bone with smooth, rounded surfaces for forming joints at both ends.

Lower leg bone
A hoofed mammal has a long, strong bone on each lower leg. This bone is really two fused, elongated foot bones.

Lower forelimb bones
Most animals have two parallel forelimb bones. They are comparable with human forearm bones (radius and ulna).

Elbow

Cloven hoof
Hoofed mammals are either two-toed (with cloven hoofs), such as deer, sheep, goats and pigs, or one-toed (with single hoofs), such as the horse.

Foot bones
All mammals except hoofed species have four or five-toed feet. Insectivores and the weasel family have five on all four feet; foxes, dogs and cats have four all round. Rodents have four front toes and five hind toes, rabbits and hares the reverse.

Shin bone
There are two bones in the lower hind leg, joined along their length. The larger one has a sharp ridge along it.

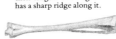

Thigh bone
One end has a large knob at one side. This fits into the socket in the pelvis.

Knee

Recognising skulls

The main ways of identifying a skull are by its size and by the type and arrangement of the teeth. Complete skulls of larger animals are often found. Those of small mammals such as mice, voles and shrews are under 1 in. (25 mm) long and are most unlikely to be discovered except in discarded bottles (the remains of trapped animals, see p. 230) or owl pellets.

Hoofed mammals

Hoofed mammals are plant eaters, so normally have no canine teeth for seizing prey. There is a big gap between the incisors (used for cutting) at the front of the jaw and the molars (used for grinding) towards the rear. Horses and pigs have incisors on both upper and lower jaws. Other hoofed mammals have a toothless pad at the front of the upper jaw.

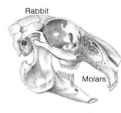

A sheep skull has incisors only on the lower jaw, which is about 8 in. (20 cm) long. It may have horns.

Roe deer

A deer skull resembles that of a sheep; only a male's has antlers. A roe deer skull is about 8 in. (20 cm) long, a red deer's about 12 in. (30 cm).

Rodents

The skull of a rodent such as a rat or squirrel has a big gap between the incisors at the front of the jaws and the molars towards the rear. Rodent incisors have a yellow or orange front surface not found in other mammals. Rats, like mice, have molars with a knobbly surface (p. 234).

A red squirrel skull is about 2¼ in. (54 mm) long at the most. A grey squirrel skull is about 2⅜ in. (60 mm) long.

Carnivores

The skulls of carnivores such as foxes and cats, which are flesh eaters, have teeth all the way along the jaw. They have large, sharp-pointed canine teeth towards the front.

A fox skull is about 5 in. (12·5 cm) long. Mink, polecat, stoat and weasel skulls are less than 2½ in. (64 mm) long.

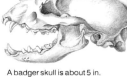

A badger skull is about 5 in. (12·5 cm) long. It is broader than a fox skull and has big, flat-topped molars. Its lower jaw will not detach from its upper jaw.

A cat skull is rounded, with short jaws, and is about 3 in. (75 mm) long. All the teeth are pointed.

Hares and rabbits

The skull of a rabbit or hare is similar to a rodent's skull but has a second pair of tiny incisors behind the grooved front pair in the upper jaw. The molar teeth have an oval surface.

Rabbit

A rabbit's skull is up to 3⅜ in. (85 mm) long. An adult hare's is longer.

Molars

Incisors

Insectivores

Like carnivores, insectivores such as shrews and hedgehogs have teeth all along the jaw, with no big gaps between incisors and molars. But unlike carnivores, they do not have big canines. Their teeth have very sharp points.

A hedgehog skull is about 2 in. (50 mm) long. Two almost horizontal lower front incisors fit into a gap between the two prominent upper incisors.

Dog

Fox

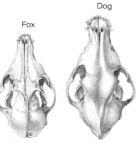

Viewed from above, a fox skull can be distinguished from a dog skull by the eyebrow ridges. In a fox they have concave pits, in a dog they are convex, giving bulging brows.

Hardy cattle of the Scottish Highlands

Hardiness was the quality that the hill farmers of north-west Scotland prized in Highland cattle. Insulated by their double coats and thick skins the cattle could winter on the hills – fortunately – for their vast, spreading horns make them difficult to house.

In an average Highland winter, snow falls on 50 days and the air can be painfully cold. But it is the rain and wind as much as the cold that the cattle have to withstand. There are few trees to give shelter, only natural hollows, and the moisture-laden winds from the Atlantic bring 80 in. (200 cm) or more of rain a year and hurl it fiercely against the slopes.

Highland cattle thrive on the rough, pallid upland grass, and scramble on sure feet to reach it. They grow slowly, taking three or four years to mature and produce the well-flavoured fine-grained lean beef for which they were traditionally reared. From the 18th century herds of Highland cattle, shod for the journey, set off in late summer down the drove roads leading south, covering about 12 miles (19 km) a day. Some began the journey by swimming across the kyles, or straits, from the islands to the mainland. These cattle, called Kyloes, were eventually fused with the larger Mainland type in the Highland breed. Lowland graziers bought the droves to fatten for the beef trade.

Today, Highland cattle are mostly crossed with Shorthorn bulls to produce faster-growing beef or cows to suckle beef animals. In parks the cattle are kept for their attractive appearance. Among them a few may still be black – the original dominant colour before selective breeders chose dun and red as the breed colours.

In summer, curlews are common on the moors and rough grassland where the cattle graze. Hooded crows and ravens often circle above, on the look-out for afterbirth or still-born calves.

The long, shaggy coat throws off rain and has a dense undercoat of soft hair that traps air against the body, giving insulation from cold. A thick forelock of long hair protects the eyes and face.

Highland cattle are Britain's hardiest breed. They winter out of doors in all weathers, but are given extra food such as hay.

Despite their wild looks, with long, sweeping horns and shaggy coats, Highland cattle are quite docile, except for cows with calves. Dun and red are the usual colours, but a few animals may be brindle, yellow or black.

Cows bear their first calf when about three and a half years old. Most calves are born between January and May. Cows will use their long horns in defence of their calves.

A mature bull may weigh a ton (1,016 kg) or more. He is aggressive and has to be handled with great care.

The horns are usually removed at birth. When present, horns are short and curve inwards.

In the last 50 years, most of Britain's milk has come from the familiar black-and-white British Friesian cow.

Black-and-white colouring

Long face

Large udder

The British Friesian is the foremost dairy cow in Britain. Its black-and-white colouring is distinctive, and it has a deep, long body with a large udder. The head is long with a straight profile. Large, cow about 1,320 lb (600 kg).

Friesians may sometimes be red and white. The Holstein (left) is leaner and more angular than the British Friesian (right) and is more of a specialist dairy cow.

British Friesian cattle *Bos* (domestic)

No other breed of cow can compare with a Friesian for high milk yield. A British Friesian cow will, on average, give just over 1,200 gallons (roughly 5,500 litres) of milk after each calving. Its relative the Holstein, bred in North America and South Africa, tops this with about 1,350 gallons (roughly 6,150 litres). To achieve this yield, the cows need lush pastures and plenty of other high-energy food.

Friesian cattle take their name from the Dutch province of Friesland. Their ancestors came from Jutland in Denmark, and spread into the Netherlands to replenish the native stocks, which were periodically devastated by cattle plague and severe flooding from the Zuider Zee. The skilled Dutch farmers developed the black-and-white cattle to produce high milk yields. They were first brought to Britain in the 17th century, and in large numbers in the late 1800s. As the Friesian spread around the world, different types developed. The large and powerful British Friesian, which provides lean beef as well as milk, is intermediate between the stocky Dutch Friesian, bred for meat and milk, and the dairy-type Canadian Holstein, which is kept solely for its milk.

Lean body

Dished profile

One of the world's outstanding dairy cows, the Jersey is lean-bodied and angular, with a well-developed udder and short, fine-boned legs. The horns turn forwards and inwards and the head has a dished (concave) profile. Colour varies from light fawn to brown and mulberry. Small, cow about 860 lb (390 kg).

Native to the Channel Islands, the Jersey has become one of the most widespread breeds in the world.

Jersey cattle *Bos* (domestic)

First brought to England from the Channel Islands in the mid-18th century, the Jersey cow became a prized possession for a small farmer. Its small size made it economical to keep and its rich milk was ideal for making butter. In fact, a pound of butter could be made with less than two gallons of Jersey milk, whereas for most other breeds the normal amount used was three gallons. Although Jerseys are not one of the highest milk producers, because of the richness of the milk they give for a long period after calving they are one of the most widely exported British breeds of cattle. They are very tolerant of heat, and have proved particularly useful in subtropical regions for cross-breeding to improve the quality of native cattle.

The Jersey's heat tolerance is inherited from Asian ancestors. The breed owes its origin to two waves of migrants. The first were prehistoric men from Asia; the second were the Vikings, who spread across Europe between the 8th and 10th centuries. The very distinct Jersey breed that evolved from the cattle brought to the island by the Asian and Viking settlers was so highly prized that in 1763 the importation of other cattle to the island for breeding was forbidden.

The bull is usually darker in colour than the cow, especially on the head and neck. He can be aggressive and unpredictable.

The head is small and delicate, with large, soft eyes. A light-coloured ring encircles the nose.

243

Large udder

Roan colouring

As with most dairy-breed bulls, the bull has an unpredictable temper and can be dangerously aggressive.

Dairy Shorthorns were common in Britain until about 50 years ago. Now they are kept mainly on small farms.

Built to produce both milk and meat, the Dairy Shorthorn has a well-developed udder and a well-fleshed body. Compared with the Beef Shorthorn, it is narrower in the back and loins and longer in the leg. Roan is the commonest colour. Large, cow about 1,210 lb (550 kg).

When horns are present, they may either turn down or tilt upwards. But most animals are dehorned at birth.

The coat may be roan, white, red or red and white. Roan (intermixed red-and-white hairs) results from mating a red with a white animal.

Dairy Shorthorn cattle *Bos* (domestic)

As more and more people moved into Britain's towns in the early 19th century, the demand for milk increased. A Yorkshireman, Thomas Bates of Kirklevington (now in Northumberland), saw this as an opportunity to develop animals of the Shorthorn breed – the most prominent cattle in Britain at that time – to produce the Dairy Shorthorn. It not only gives a high milk yield – an average of more than 1,000 gallons (roughly 4,700 litres) after each calving – but produces high-quality beef. At the same time it is an economical feeder, thriving mainly on the normal food available on a mixed farm, with some extra rations. It spread to all parts of Britain and was the most numerous breed until superseded by the British Friesian.

Shorthorn cattle existed in north-east England as early as the mid-1500s, descended mainly from Dutch short-horned cattle but with some Scandinavian influence. The breed was improved in the late 1700s by the brothers Charles and Robert Colling of Darlington. In the 19th century breeders developed it into several different types, the Dairy Shorthorn being the most successful. Other types include the Beef Shorthorn, Lincoln Red, Whitebred Shorthorn and Northern Dairy Shorthorn.

White body

Black muzzle

Short legs

Naturally hornless, the British White has a high, pointed poll (or crown). Its black-rimmed eyelids and black ears and muzzle contrast starkly with the smooth white of its face.

The British White is a minority breed kept mainly for its colour markings. Small numbers have been exported.

The British White has very distinctive markings. Its coat is smooth and white but it has black points – that is, ears, muzzle, feet and teats. It is stocky and short-legged and the bull has a broad face. Medium, bull about 1,760 lb (800 kg).

The teats are usually black, but may be pink on some animals. Many British Whites have a black patch on the back of each thigh.

Calves are usually reared by their mothers rather than hand-reared. The cows are attentive and very protective mothers.

British White cattle *Bos* (domestic)

Whalley Abbey, near Blackburn, Lancashire, is the first known home of a herd of British White cattle, which were probably once kept by the Cistercian monks of the abbey. There is a record of the dispersal of the herd in 1697. The breed originated in the 16th century as a result of crossing a polled (hornless) bull, probably of Scandinavian origin, with native white cattle. White Park cattle, a rare ancient breed, have similar markings to the British White but are horned. British Whites were once thought to be a variant of the White Park breed, but in fact are not closely related. Some of the Whalley cattle went to the Somerford estate of Sir Walter Shakerly in Cheshire, and the survival and type of the breed was largely controlled by Sir Walter's wishes. Improvements were made by using bulls of other breeds, including the Fjallras from Sweden.

Originally, British Whites were kept as meat and milk animals, chiefly for milk. But their milk yield is not high enough to compete with modern dairy breeds, and they are now considered a beef breed. Although their meat output cannot compare with specialist beef breeds, their yield in proportion to their food requirements makes them a most efficient breed.

245

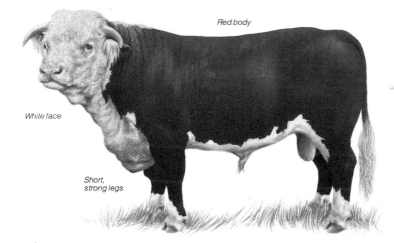

Red body

White face

Short,
strong legs

The cow, like the bull, is usually docile. It is kept for bearing and suckling beef calves, not for its milk yield, which is low.

The white face of the Hereford is a dominant characteristic, passed on to its progeny even in cross-breeding.

Hereford bulls are often mated with Friesian cows to produce white-faced black-and-white calves. The cross-breeds are popular early maturing beef animals.

A world-famous beef breed, the Hereford is distinguished by its white face and deep red body. Its chest, lower legs, tail tassel and underparts are also white. The body is deep and thickset and the legs short and strong-boned. Medium, bull about 1,870 lb (850 kg), cow about 1,100 lb (500 kg).

The Hereford's horns may grow almost horizontally or curve markedly downwards. Some animals are born without horns, and the number is increasing.

Hereford cattle *Bos* (domestic)

Hereford cattle are found on the great ranges of the North American West and the pampas lands of South America, on the scorched South African veldt and in the Australian outback. The well known white-faced, red-bodied cattle are the most widely exported British breed, chosen because they do well in rich pastures or on arid plains, in tropical heat or Arctic cold.

Herefords were originally meat, milk and draught animals, renowned in their native Herefordshire for their hardiness. Dutch cattle are among their ancestors, and they share their white head with the Dutch Blaarkop breed. Their fame spread, and by the 18th century Herefords were considered the foremost cattle in Britain. Their development as a modern breed was chiefly the work of an 18th-century breeder, Benjamin Tomkins, and his family. They concentrated particularly on selecting early maturing animals. In the early 19th century there were grey, roan and mottled-face Herefords, but the typical red body and white face was selected as the official colour of the breed during the 1800s. In Britain today, the Hereford is used mainly to mate with dairy cows to produce beef calves and suckler cows for rearing beef calves.

Shaggy coat

Sturdy legs

The Galloway will thrive and produce good quality beef on poor, upland grass where other breeds could not survive.

A hardy beef breed, the Galloway is protected by a shaggy, water-repellent coat with a dense, short undercoat for warmth. It is compact and deep-bodied with sturdy legs. The coat colour is commonly black but there are variants, chiefly dun. Medium, bull 1,700 lb (771 kg), cow 1,056 lb (480 kg).

The short, broad head is naturally hornless and flat between the ears, not pointed like an Aberdeen Angus head.

The cow is usually less docile than the bull and is aggressive in defence of her calf. Beef cows suckle their calves until they are about nine months old.

Galloway cattle *Bos* (domestic)

During the 17th and 18th centuries, as many as 30,000 Galloway cattle a year made the long trek from Scotland down the old drove roads to the lush pastures of the Midlands and East Anglia. There the three or four-year-old animals were fattened up for Smithfield meat market in London. The Galloway is one of the old, black, Celtic breeds from the western and northern fringes of Britain. Some of its ancestors may have been Scandinavian cattle, probably brought over by the Vikings, because all Galloways are hornless and some have dun colouring – both features of some Scandinavian cattle.

Galloways have always been beef cattle. They are rugged and well able to withstand the rigorous climate of their native uplands, and to thrive on the sparse upland grass. Most are still found in the south-west corner of Scotland and just over the border in northern England. They are reared for beef on a small scale, but are rather slow growing. Today their main value is for cross-breeding. Cows are mated with White Shorthorn bulls to produce crossbred Blue Grey cows, popular for breeding hardy but quicker-maturing beef calves and sometimes as foster mothers for suckling calves in beef herds.

247

Cattle for milk and beef

British cattle have been bred as either dairy animals – British Friesians and Jerseys (pp. 242–3), for example – or beef animals, as with Herefords and Galloways (pp. 246–7). With breeding controlled by artificial insemination, the most profitable breeds have spread all over the country and ousted many local breeds. Less-common breeds are still valued for their varied qualities, and may be kept by small farmers or used in cross-breeding.

Simmental

A large Swiss breed with a well-developed dewlap. Widespread on the Continent, it has been kept in Britain for beef and milk since 1970. Colour: from yellowy to deep red; head, legs, tail-tip white.

Dexter

An extremely hardy dwarf breed from Ireland, now kept mostly in England, the Dexter will thrive on poor grazing and give rich milk and fine-grained beef. Colour: usually black, may be red or dun.

Beef Shorthorn

Short-legged, muscular and compact, the Beef Shorthorn matures quickly in varied conditions and has been widely exported, especially to Australia and Argentina. Colour: usually red, white or roan.

Sussex

Originally bred to pull carts or ploughs, deep red Sussex cattle thrive on a wide range of grasses and have been widely exported, particularly to South Africa. They are early maturing and yield lean meat.

Aberdeen Angus

The docile, hornless Aberdeen Angus is short-legged, stocky and muscular and a supreme beef animal. Bulls are used to improve the beef quality of cross-bred calves. Colour: black, occasionally red.

Lincoln Red

A strong, long-bodied breed from the rich Lincolnshire pastures, now kept mostly for beef. Many are hornless. Cows are good mothers and are used for producing cross-bred beef calves.

Devon
Hardy, agile and deep-chested Devons were once used as draught animals. They thrive well on poor grazing and are kept for beef, mainly on small farms. The thick red hair is often curly.

Guernsey
The lean, long-faced Guernsey gives rich, yellowish milk suitable for cream and butter. It is noted for its longevity. Colour: usually red with white, sometimes brindle or black and white.

Charolais
A heavy beef animal from France, first kept in Britain in 1961. Exceptionally lean and quick-growing, it is much used in cross-breeding to improve beef yields. Colour: always creamy-white.

Welsh Black
An ancient breed derived from Celtic stock, the Welsh Black is a self-reliant forager able to winter out of doors. Its milk is high in butterfat and it matures slowly to produce lean beef.

South Devon
A hardy, docile breed with curly red hair. It comes from the lush Devon lowlands and is one of the largest native cattle. Its rich milk is used for clotted cream and it provides lean beef.

Ayrshire
A hardy, small-boned dairy animal second only to the Friesian in milk output. The milk is suitable for cheese-making. Colour: normally brown on white, sometimes black on white.

Blue Albion

Bred in Derbyshire about 1920–40, the Blue Albion gained its blue roan colour from the cross-breeding of Welsh Black cattle with Shorthorns. Blue cattle do not breed true for colour. It is unlikely that any pure-bred Blue Albion remain.

Belted Galloway

Cross-breeding with Dutch belted cattle produced the belted type of Galloway, probably as a decorative animal for estate farms. Registered as a separate breed in 1921, it has the hardiness and beef quality of all Galloways.

Gloucester

True Double Gloucester cheese was made from the milk of Gloucesters, developed as a dairy breed 200 years ago. The white tail and 'finching' (back stripe) are characteristic of the breed.

Kerry

A small dairy cow dominant in western Ireland 100 years ago. It is hardy, thrives on scanty food and is probably the breed most like the earliest British cattle of about 2000 BC.

Irish Moyled

The rarest cattle breed in Britain, the Moyled is a milk and beef breed that takes its name from an Irish word meaning 'little mound' – a reference to its domed head. It is hornless, probably from Viking cattle in its ancestry.

Longhorn

Longhorns, which originated in the Craven district of Yorkshire, were among the most popular British cattle until about 1800. They were sturdy draught animals, gave rich milk for butter and made good beef. Shorthorns superseded them.

Rare cattle in Britain

As a few breeds increasingly dominate Britain's cattle industry, some other breeds have become so rare that their survival is endangered, and they are maintained only by conservation projects. Once-common dual-purpose breeds such as the Dairy Shorthorn (p. 244) have lost popularity, and some other breeds, the British White (p. 245) for example, have never been numerous. A few breeds that are in a minority in Britain today are outstanding foreign breeds only recently introduced; these may increase in numbers.

White Galloway

The breed originated in the late 19th century when black Galloway cows were cross-bred with a White Park bull. White Galloways were registered as a separate breed in 1977. Like the black Galloways, they are hardy beef animals.

White Park

An ancient British breed that survived on estates at Chillingham (Northumberland), Dynevor (South Wales) and Cadzow (East Lothian). White Park cattle have been sacrificed by Druids, hunted and used as draught animals. Today they are used to sire cross-bred beef calves.

Romagnola

A dark-skinned, black-tongued breed that has proved adaptable to weather extremes since first brought to Britain from its native central Italy in 1974. Its fine-grained, quick-maturing beef is well flavoured.

Shetland

An ancient breed that tolerates the harsh conditions and poor grass of the Shetland Islands. It has been extensively used for cross-breeding because of its thrift and mothering qualities; pure-breds are rare.

Limousin

A native of central France, the large, well-muscled Limousin gives high-quality lean beef. There are now more than 6,000 pure-bred Limousins in Britain; the first ones were brought here in 1971.

Red Poll

A milk and meat East Anglian breed that is docile, hornless and cheap to feed. It spread to Europe and America, but its numbers have dwindled with the increase of Friesians during this century.

Straight ears

Level back

Long, curled wool

Wensleydale lambs grow very rapidly. The sheep are usually kept in small flocks in fields near the farmstead.

The docile Wensleydale has a broad, level back and long, lustrous wool that curls into ringlets. Neither sex has horns. The skin of the face and ears has dark blue pigmentation, and the ears are long and carried just above the horizontal. Large, ram 287 lb (130 kg) or more.

A newly shorn animal can usually be distinguished by skin colour and ear position. Wool grows on the forehead, the cheeks, and the legs as far down as the fetlocks.

A few Wensleydale sheep are black, the darkness of the skin pigmentation extending over the whole body.

Although Wensleydale ewes are good mothers, their lambs sometimes have a difficult time finding the udder because of the long wool. A ewe weighs 170–180 lb (77–82 kg).

Wensleydale sheep *Ovis* (domestic)

One of the largest British sheep, the Wensleydale is mostly confined to the Yorkshire Dales and at one time was almost extinct. Now the breed is becoming increasingly popular because of the special quality of its wool, as well as its ability to sire heavyweight lambs. It yields a very heavy fleece, about 14 lb (6·4 kg) on average, and the long ringlets of its lustrous wool are free of coarse fibres. The average length of fibres is 12 in. (30 cm). Wensleydale wool is used for lining jackets and other garments, and also for tapestry weaving.

The Wensleydale breed was developed following the birth of a ram named Blue Cap in North Yorkshire in 1839. Blue Cap was a cross between a Leicester ram and a Teeswater ewe, and was distinguished by the deep blue colouring of his head. This characteristic was transmitted to his descendants, which became the modern Wensleydale breed. It has been found that the deep blue pigmentation of the skin protects the animal against sunburn, and has enabled Wensleydale sheep to be exported to tropical countries. The Wensleydale has also been used for selective breeding in Europe, and was one of the parent breeds of the French breed Bleu du Maine.

The bold, strong head is black with some white markings and is heavily horned. Rams particularly have a high-bridged nose.

Long coarse wool

Short legs

Black face

The Scottish Blackface is a hardy and vigorous mountain breed that will dig through snow to find grass. Its wool is long and coarse and it has a robust body and short, mottled legs. Lambs are born in late March or April. Medium, ewe 130 lb (59 kg), ram bigger.

Rams often fight, especially during the breeding season in October or November. Their curved, spiralled horns grow flat from the top of the head.

In the north, shearing usually takes place in July or August. The fleece generally weighs about 5½ lb (2·5 kg). Wool does not grow on the sheep's face or lower legs.

Ewes have lighter, less-curved horns than rams. They will fight in defence of their lambs.

Scottish Blackface sheep *Ovis* (domestic)

The most popular breed of sheep in Britain, the Scottish Blackface is found throughout most of the Scottish Highlands and also in many other areas, including Ireland, Dartmoor and East Anglia. Because of its heavy fleece, which hangs almost to the ground when full, the sheep can thrive in a harsh environment exposed to strong winds and high rainfall. It is an active animal with the energy necessary to forage widely in search of food, and both sexes are horned and aggressive – well able to defend the lambs against predators.

Black-faced mountain sheep may originally have been brought to Britain by the Vikings. The beginnings of the Scottish Blackface breed can be traced to the Pennine Hills of northern England, but the sheep spread through Scotland in the 17th and 18th centuries. Along with the Cheviot, it was the sheep of the Highland Clearances, when families were turned off the land to make way for flocks. Various different types of the Scottish Blackface have developed; the Lanark type has a heavier fleece than the Newton Stewart type, and the Boreray is an earlier kind with a light fleece. The wool – long, strong and coarse – is used for stuffing mattresses and making carpets.

Short wool

The ram's head is broad and strong, the nose slightly high-bridged. During courtship he may hold up his head and turn back his top lip.

The Suffolk is a lowland breed originating in the pastures of East Anglia, but now widespread.

The Suffolk is a broad-bodied sheep with short, fine, tight wool. It has short black legs and a black face with long ears carried horizontally and curving backwards at the tips. Both sexes are hornless. Large, ram up to 298 lb (135 kg).

Black face

The breed is noted for its broad hindquarters with large thigh muscles. The tail is cut short.

Ewes weigh about 180 lb (82 kg) and have longer, narrower heads than rams. Lambs are born as early as November or December.

Suffolk sheep *Ovis* (domestic)

Ideally suited for meat production, the Suffolk is a large, powerful sheep with thick bones, a strong, well-developed body, very little fat and thick muscling on the hindquarters. The breed was developed in the late 18th century in East Anglia, by crossing Norfolk Horn ewes with Southdown rams from Sussex. Norfolk Horn sheep were well adapted to survive on the sandy heaths of the East Anglian Breckland, but as these were enclosed and cultivated the breed's lean hardiness was no longer needed. The Suffolk, bred for the new fertile pastures, replaced them. Today the Norfolk Horn is almost extinct and the Southdown a minority breed, but the Suffolk has become internationally popular.

In a cold area with low rainfall such as the east coast, the Suffolk's tight fleece of fine wool is ideal. Because the breed has proved to be adaptable, it has been exported to many parts of the world with varying climates. The Suffolk's black head and legs, which are free from wool, are inherited from the Norfolk Horn. The tail is cropped at birth because lowland sheep, which graze on rich pasture, rapidly become soiled if the tail is long and woolly. Lambs often have black flecks in their fleeces.

Spiralled horns

Hairy coat

Large, spiralled horns with corrugated growth rings distinguish the powerful Wiltshire Horn ram. The breed has a matted, hairy coat, and smooth white hair on legs and face. Large, ram up to 287 lb (130 kg), ewe 175 lb (79 kg).

The rams are bold and aggressive and will fight during the breeding season (October–November) They have strongly shaped heads with high-bridged noses.

The Wiltshire Horn sheds its hairy coat in summer. It is the only British sheep breed that does not produce wool.

Ewes have shorter, smaller horns than rams. The lambs are very active.

The Wiltshire Horn is a robust and well-fleshed breed. Rams are used to produce crossbred meat sheep.

Wiltshire Horn sheep *Ovis* (domestic)

The Wiltshire Horn sheep, one of a group of white-faced breeds native to south-west England, died out in Wiltshire about 1820. However, the sheep had previously become established in Northamptonshire and Anglesey, and these are still the main breeding centres.

Unlike any other modern British sheep, the Wiltshire Horn does not produce wool. Originally it had a light fleece of thin, fine wool, but the normal growth of its fleece has been deliberately suppressed by selective breeding. Today the Wiltshire Horn's short, hairy, matted coat peels off as the sheep gets fatter on the rich grass of late spring and early summer. The breeders claim that the sheep compensates for its lack of wool by laying down extra flesh, and also that many of the skin parasites associated with long wool are avoided.

Both the ewes and rams have horns, those of the rams being particularly large and strong. On all strong-horned sheep, the horns grow rapidly during spring and summer and very little in winter. These phases of growth are shown by the size of the ridges on the horns, and can be used to discover an animal's age. The natural life-span of a sheep is about seven years.

Heavy
fleece

Some Milksheep have a
bare area round the tail, so
their tails may be left long
rather than cropped for
reasons of hygiene.

Long
white
legs

Modern British Milksheep help to recall
the Middle Ages, when sheep were often
kept as dairy animals.

A long-bodied breed with long white
legs and a heavy white fleece, bred for
both milk and wool. Despite its alert
appearance it is very docile. Both rams
and ewes are hornless. The face may
be slightly speckled. Medium, ewe
170 lb (77 kg), ram 225 lb (102 kg).

Milksheep have pink or partly
pink muzzles and large, long
ears with little hair on the insides.
Their faces are fairly long and not
very broad, and they have a
straight profile.

The ewes are good mothers and
very protective of their lambs.
They can yield 140 gallons
(about 635 litres) of milk yearly.

British Milksheep *Ovis* (domestic)

Most breeds of sheep produce one or two lambs at a time, but
the British Milksheep averages three, and is also able to yield
sufficient milk to rear all of them. A relatively new breed, it was
developed in England in the 1960s and 1970s and is now found
throughout Britain, particularly in the south-west. It is popular
because of its high productivity, and is also suitable for producing milk for the small British market in ewes' milk and cheese.

Five breeds were blended to produce the Milksheep. One of
them was the Fries Melkschaap of the Netherlands, noted for its
milk yield. Another was the Poll Dorset which (with the related
Dorset Horn) is one of the only two British breeds able to
produce lambs at any time of the year. Also involved were the
Bluefaced Leicester, noted for its high growth rate, and the
Lleyn, a good all-round Welsh breed. The fifth breed concerned
was the Prolific, an intermediate breed now extinct.

The British Milksheep is built for easy lambing, and its large,
strong udder allows a large milk capacity. The fleece averages
10 lb (4·5 kg) and the wool is semi-lustrous, white and of high
quality, making it suitable for worsted – a fine, soft woollen
yarn – and hand-spinning.

Black-and-white fleece

White blaze

Sheep of the recently improved type are bigger, with shorter legs and a thicker body. On any newly shorn Jacob sheep, the black wool looks jet black.

Jacob sheep are now kept for their wool, on a minor scale, and flocks may be seen in many areas of Britain.

The only sheep in Britain with a black-and-white fleece. The face usually has a white blaze with black cheeks and eye surrounds, and sometimes a partly black muzzle. Generally, a Jacob has either two or four horns; ewes have smaller, shorter horns than rams. Small, ewe 99 lb (45 kg), ram bigger.

A four-horned ram has one strong, slightly curved pair growing upwards, and a smaller pair growing sideways, often curling towards the face. A two-horned ram has strong, spiralled horns.

The black wool is longer than the white; on a full fleece it bleaches to a dark brown. Typical, unimproved Jacob sheep are small but long-legged.

Jacob sheep *Ovis* (domestic)

Jacob sheep take their name from their spotted fleece. The story of Jacob in the Book of Genesis tells how, when Jacob wanted to leave his employer Laban to set up on his own, it was agreed that his wages should be (according to an early translation) all the spotted and speckled animals from Laban's flocks and herds. But the story is confused by a later translation that refers to the sheep as brown rather than spotted.

Many Jacob sheep have one or two pairs of horns – a few have three pairs. They are descended from a Spanish breed, the same sheep that were taken to North America by the Spanish *conquistadores* and were the ancestors of the multi-horned sheep now farmed by the Navaho Indians.

In Britain, Jacob sheep have traditionally been kept as a novelty attraction in estates and parks. Recently they have gained a wider acceptance as a breed of special interest, and an improved type has a minor role in the wool industry. The wool is used for hand-spinning to produce naturally coloured fabrics. Although coloured wool is a hindrance in the mechanised processing and dyeing of wool, it has extra value in hand-spinning and weaving.

257

Longwools and crossing breeds

The Romans brought to Britain the longwool sheep from which lowland pasture flocks were developed. These were the sheep on which much of Britain's medieval wool and cloth trade was founded. Today there are two types, those used mainly for their wool, including the Wensleydale (p. 252), and those popular for cross-breeding, such as the British Milksheep (p. 256).

Dartmoor
Closely related to the Devon and Cornwall Longwool, the Dartmoor is smaller and has a better-quality fleece which rivals that of the rare Lincoln Longwool in weight. Rams may occasionally have horns.

Romney
Isolated on Romney Marsh in Kent, the breed is little different from medieval lowland long-wool sheep, being hardy and well fleeced but not prolific. The breed is hornless.

Devon and Cornwall Longwool
A large and heavy-boned sheep with a long, lustrous, curled fleece, but not a prolific breeder. Descended from local shortwool types, it has black spots on its ears. Neither sex has horns.

Border Leicester
Bred in the Scottish Borders from Leicester Longwools and Cheviots, the breed is valued for its rams. They are mated with slow-growing hill ewes to give prolific and quick-maturing cross-bred lambs. Neither sex has horns.

Colbred
A hornless 20th-century breed developed for crossing with hill ewes to give early maturity and high milk yield. It is long in the leg and body and has pink nostrils and a silky fleece.

Bluefaced Leicester
A large, long, thin-fleeced, hornless sheep bred from a dark strain of Border Leicester and even more prolific. It is not hardy, but rams are mated with hill sheep to produce prolific cross-bred ewes.

Down and grassland breeds

Since selective breeding began in the mid-18th century, breeds have been developed to be prolific and produce meaty carcasses of lamb quickly. Sheep bred for lowland pastures are larger and earlier maturing than hardier mountain breeds. The Down breeds, including the Suffolk and the Wiltshire Horn (pp. 254–5), are noted for the high quality of their meat.

Black Welsh Mountain
One of several small and hardy mountain breeds developed from the tan-faced native Welsh sheep. It is now kept mainly in parkland flocks for its soft black wool. Rams have horns.

Clun Forest
A hornless grassland sheep bred in the West Midlands from Shropshire downland and Welsh hill sheep. It is adaptable, forages well and is a prolific producer of quick-maturing lambs.

Hampshire Down
A short-legged, short-bodied, hornless sheep bred from the local sheep and the Southdown as a meat breed to mature quickly on chalkland grass. The ewes have a long breeding season.

Dorset Down
A high-quality meat sheep, the rams being used to breed meaty, quick-maturing cross-bred lambs. Both sexes are hornless. The wool is used for fine knitwear and in paper for banknotes.

Devon Closewool
A medium-sized, self-reliant grassland sheep bred from the Devon Longwool and the thickset, densely fleeced Exmoor Horn hill sheep. Both sexes are hornless.

Kerry Hill
A hardy, active grassland sheep that produces a soft fleece. It is descended from sheep of the Eppynt Mountains in Wales, and has a black muzzle and black spotting on the face and legs. Rams occasionally have small horns.

Dorset Horn
A pink-muzzled grassland sheep that is, along with its hornless counterpart the Poll Dorset, the only British breed that lambs at any time of the year, so bringing a high price for out-of-season meat. The fine-textured wool is used for socks and dress fabrics. Both sexes have horns.

Hill and mountain breeds

The original sheep of Britain were small and hardy with tan faces, and were valued as much for their milk as for their meat, skin and wool. Some tough hill breeds are descended from them. Hill and mountain breeds, such as the Scottish Blackface (p. 253), can survive in cold, wet conditions on coarse hill grass. They are usually leaner and less prolific than lowland breeds.

Derbyshire Gritstone
A Peak District breed developed from black-faced hill sheep and Leicesters. It is the biggest of the hill sheep and is crossed with Down breeds to produce meat lambs. The breed is hornless. Its wool is used for socks.

Dalesbred
Its long, crimped fleece with a short, dense undercoat protects the sheep from Pennine winters. It has a white patch on each side of its black, grey-muzzled face. Rams and ewes are both horned. The wool is used for carpets.

Radnor
Very little changed from the old tan-faced breed of mid-Wales, the sheep is halfway between the hill and grassland types. Its coarse fleece is used in woven fabrics. Rams have stout, curled horns.

Lonk
Because it has a short but dense fleece and long legs free of wool, the Lonk can move with great agility on its high Yorkshire moorlands. Legs and face are patterned black and white. Rams' horns have sweeping curves; ewes' horns are less curved.

Swaledale
The white muzzle and black face are distinctive. Like the Dalesbred, the sheep has a dense undercoat. It is the most numerous breed in the northern counties and is widely used for cross-breeding. Both rams and ewes have horns, which are used for making ornate carved crooks.

Rough Fell
A placid, persistent forager, the sheep is nimble enough to reach the inaccessible nooks of the high, slaty Pennine fells. Both rams and ewes are horned. The straight, coarse wool hides the feet in full fleece, and is used for carpets and rough fabrics.

Exmoor Horn

White-faced horned sheep were the native type of south-west England. The Exmoor Horn is little changed from them, but has a better-quality fleece and matures quicker. It is thickset and broad-faced and not particularly hardy, but its dense fleece throws off rain. Both sexes have horns.

Beulah Specklefaced

One of the Welsh Mountain breeds, a hardy but prolific sheep native to the hills of South Wales. The speckled face may be inherited through cross-breeding from the Derbyshire Gritstone. Both sexes are hornless.

Welsh Mountain

The most numerous of the several related breeds developed from the original mountain sheep of Wales, and found in the north around Aberystwyth. Small with rather a coarse fleece, the sheep is useful for grazing rough land. The rams have curling horns.

Cheviot

The modern Cheviot – bred from the tan-faced, small-boned, horned original – is a medium-sized producer of small meat lambs and is best suited to the grassy border hills rather than rougher mountain areas. The white, high-bridged nose is characteristic. Rams occasionally have horns.

Whitefaced Dartmoor

A local breed developed from the original horned, white-faced West Country sheep. Its long, lustrous fleece is unusual in hill sheep. Rams occasionally have horns, a throwback to the original breed.

North Country Cheviot

A larger version of the Cheviot developed in the north of Scotland. It has the same white, high-bridged nose, soft wool and clean legs as the Cheviot, and is usually also hornless. It produces high-quality meat.

Leicester Longwool

A large sheep with a heavy, curled fleece. It is bred mainly in the Yorkshire Wolds for its late-maturing heavyweight lambs. Neither sex has horns.

Manx Loghtan

A primitive breed from the Isle of Man. Both ewes and rams usually have two pairs of horns. The name comes from *loghtan*, the Manx for mousy brown, which describes the sheep's colour. The wool is used undyed for knitwear.

Portland

One of Britain's original tan-faced breeds. Although born with a tan fleece, the sheep soon becomes white or grey with some coarse tan fibres in the short, fine wool. The Portland is small, and both sexes have horns. The lambs give well-flavoured, fine-textured meat.

Whitefaced Woodland

A large hill breed of south-west Yorkshire and Derbyshire that is sometimes called the Penistone. Its pink nostrils and long tail are characteristic. Ewes and rams have horns. The ram is used to breed size and lambing vigour into ewes of black-faced hill breeds.

Cotswold

A large breed known since medieval times, when the Cotswolds were a centre of the wool trade. Its heavy, curled fleece, often matted, has fibres up to 12 in. (30 cm) long. It is slow-growing and tends to put on fat. Both rams and ewes are hornless.

North Ronaldsay

A small, hardy, fine-boned primitive breed that grazes almost entirely on seaweed round the coast of North Ronaldsay in the Orkneys. The fleece colour varies, ranging from white through to black. Rams have horns.

Rare breeds

There are two kinds of rare sheep breeds. Some, such as the North Ronaldsay, are primitive native sheep whose qualities have not been improved by selective breeding, usually because they have survived on isolated islands. Other breeds, such as the Southdown, are rare because they have been superseded by breeds with more profitable qualities. Jacob sheep (p. 257) are an ancient breed kept mainly for their appearance.

Oxford Down

One of Britain's biggest sheep, bred from the Cotswold and the Hampshire Down. It is a meat breed that produces high-quality quick-growing lambs. The wool is used for knitting yarns. Neither sex has horns.

Shetland

A small, very hardy sheep prized for its soft, fine wool of very high quality that is used in Shetland knitwear. Fleeces are shed in summer if not plucked or shorn, and weigh only about 3 lb (1·3 kg). They may be white, black, grey, brown or rust. The rams have small horns.

Southdown

Bred to be a supreme meat animal, the sheep is hornless, small-boned, fleshy and quick to fatten. The New Zealand lamb industry was based on Southdowns, but now only a small number of breeding ewes remain in Britain. Other breeds descended from the Southdown have eclipsed it.

Teeswater

A large, hornless sheep confined mainly to north-east England. It is very prolific, often producing triplet lambs, and is used chiefly to transmit this characteristic through cross-breeding. The fine, long-fibred fleece weighs up to 15 lb (6·8 kg).

Ryeland

A sheep bred in the 19th century from the original Hereford, which was renowned for its soft, dense wool. The lightweight fleece is used for tweeds and knitting wools. Ryeland lambs give high-quality meat. Both sexes are hornless.

Hebridean

A black sheep related to the Manx Loghtan. Like the Loghtan, it is larger and longer-tailed than other primitive breeds. The silky wool is used undyed for knitwear. Ewes may be hornless, or, like the rams, have two, four, or occasionally six, horns.

Lincoln Longwool

The sheep's face and legs are almost hidden by its long, heavy fleece which has fibres about 13½ in. (34 cm) long. The wool is used for specialist yarns. Not many breeding ewes are left. The rams are used to produce late-maturing cross-bred lambs. Both sexes are hornless.

Herdwick sheep on the Lakeland fells

The fells of the Lake District owe several of their characteristic features to the local Herdwick sheep, now a minority breed. Britain's hardiest sheep breed, the Herdwicks are out on the slopes in winter and summer. They feed persistently on seedlings, leaves and rough bent grass, so where they graze few trees survive to interrupt the wide, green views. Nor are there many walls or fences to break the landscape, because Herdwicks, like some other hill breeds, are noted for their 'hefting' instinct; each sheep will stay in the area where it was raised, or heafed, by its mother.

A Herdwick's fleece has fibres up to 7 in. (18 cm) long and contains many kemps, or coarse hairs, that shed the rain – essential in an area of high rainfall. The kemps will not absorb dye, but tweed in shades of grey and black is made by mixing fleeces from animals of different ages.

The origin of the breed is not known, but it has probably been kept in Britain for centuries. Cistercian monks at Furness Abbey are known to have kept 'Herdwycks' in the Middle Ages. Today the sheep are usually sold to lowland farmers to be fattened for meat. They cannot compete with specialist breeds and numbers have declined.

Curlews call above the grassy slopes where the sheep graze, and buzzards wheel over the high fells. Sharp-eyed enough to spot a beetle on the ground, buzzards will soon find and feed on a dead sheep or stillborn lamb.

Ash and willow trees may be seen on hillsides. The sheep readily eat ash leaves, which were once used as extra winter fodder.

Lambs are born black. As they grow they gradually get paler. Young sheep are light brown, older adults pale grey.

Hardy Herdwick sheep are short-legged, strong-boned and stockily built, and can thrive on the poor upland grass. Ewes (females) are hornless.

One ram usually runs with each flock of ewes. A ram has curled horns and a heavy mane.

Dished profile

Long body

A heavily built, long-bodied bacon pig, the Large White has strong, straight legs and white skin, occasionally with darker patches. Its head has a slightly concave (dished) profile. Large, boar up to 1,122 lb (510 kg) when fat.

Hot sun can give Large White pigs sunburn, but they may be let out to root in shady fields on cool, mild days.

Sows are prolific. They produce two litters a year, often with ten or more piglets, and suckle them for up to five weeks.

The ears are erect, with curved tips and a fringe of silky white hairs.

Large Whites may be seen out of doors during the day on sheltered land, but they cannot tolerate cold and damp and are more often housed indoors.

Large White pig *Sus* (domestic)

Pigs were for centuries commonly kept in cottage back yards all over Britain, and were bountiful providers for the household. They fed on whey and skim milk from the butter churn and cheese vat, undersized root vegetables and various household scraps. These they converted into all sorts of meals: offal to eat right away, a leg of pork for Christmas, a ham for winter, sides of bacon for breakfast and high tea, bones for broth, trotters to set brawn, cheeks for pickling, scraps for potted meat, blood for black pudding, lard for home baking and suet for puddings.

Such a pig was the Large White, evolved from a mixture of the native pigs of north-western Europe and imported Asian pigs. It was developed mainly by Joseph Tuley, a weaver from Keighley in West Yorkshire, and attracted attention at the Royal Show at Windsor in 1851. Since then it has become the most popular pig in Britain, and has been exported to many countries, including Australia and the United States. Outside Britain it is known as the Yorkshire pig. The breed has been selectively improved to become a high-quality bacon pig, which requires a long, lean carcass. The sows produce large litters but tend to be temperamental and are not always good mothers.

White saddle

Drooping ears almost hide the straight profile. The jowl is moderately heavy.

A heavy pig with a deep body, short legs and well-developed hams (buttocks and thighs). The Saddleback name is derived from the white band, varying in width, that encircles the animal's foreparts. Large, boar up to 1,100 lb (500 kg) when fat.

The Saddleback has pigmented skin that protects it from sunburn, and can be kept outdoors in the summer.

Because it is hardy enough to withstand cold and damp, the Saddleback is widely kept in outdoor pig-rearing systems.

Wessex

Essex

The original Saddleback breeds were the white-belted Wessex and Essex. The Essex also had white on the tail and hind feet.

The even-tempered sows are excellent mothers with 12–14 teats and ample milk for rearing strong piglets.

British Saddleback pig *Sus* (domestic)

The hardy British Saddleback is well able to withstand the cold and damp of outdoor life, and is not affected by sunburn to the same extent as most white-skinned pigs. It has been known as the British Saddleback only since 1967, and is an amalgamation of the old Essex and Wessex Saddleback breeds. The saddleback characteristic was probably derived from a belted Italian breed introduced from Siena.

The short, sturdy Saddleback is a dual-purpose animal, suitable for both pork and bacon, and in the past its hardiness and good-quality meat made it an excellent pig for smallholders and cottagers. Today, its bacon cannot compete with that of the specialist bacon breeds such as the Large White and Landrace, which are intensively reared indoors. But Saddlebacks are particularly good mothers, producing large, strong litters that grow rapidly, and are commonly used for cross-breeding. The sows are mated with Large White or Landrace boars to produce hybrid daughters for use as breeding sows. These cross-bred offspring, known as blue pigs, inherit some of their mother's maternal qualities and hardiness and some of their father's high quality. They make good fattening pigs for mixed farms.

Breeds of pig kept in Britain

Modern breeds of pig in Britain are descended from native wild boar and imported Chinese pigs. The Chinese pigs were brought in during the late 18th century, either directly from Asia or indirectly through Italy. More recently, breeds have been imported from North America and mainland Europe. Pigs are reared either for bacon, for example the Large White (p. 266), or pork. Some dual-purpose breeds, such as the British Saddleback (p. 267), tend to be used more for cross-breeding.

Gloucester Old Spot
A large, hardy pig, now rare, that is reared for pork. It was often kept in West Country orchards where it fed on grass and windfall apples. It is not as spotted as it used to be.

Large Black
A large, lop-eared pig with lightly pigmented skin beneath its thick black hair. It is kept for both pork and bacon, does well on poor feed, and is suited to outdoor conditions.

Welsh
A typical pig of the Landrace type, but hardier than the Scandinavian breeds and suitable for keeping out of doors. It matures early as a lean bacon pig and is also widely used in cross-breeding.

Berkshire
A small, early maturing pork pig that inherits its short, broad body and small dished (concave) face from Chinese pigs imported in the 18th and 19th centuries.

Landrace
Pigs of this fairly large, lop-eared type, long in the face and body, are native to Britain and Scandinavia. The Scandinavian breeds have been developed as prime bacon pigs and need to be kept indoors.

Chester White

A prolific pig bred in Chester County, Pennsylvania, USA, in the 19th century, mainly for bacon. It is a cross between the Large White and two breeds now extinct, the Lincolnshire Curly Coat and the Cumberland, and has white skin with blue freckles.

British Lop

One of the two British Landrace types, along with the Welsh. It used to be known as the Long White National Lop-eared and was bred in the south-west. It is a hardy pig that can be reared outdoors.

Middle White

A small pig bred to combine the small bones of the now-extinct Small White and the lean, early maturing carcass of the Large White. It is reared for pork and used in cross-breeding.

Tamworth

The Tamworth inherits its long snout from the wild boar and its colouring from an imported West Indian pig. It is hardy and resistant to sunburn. Bred in the Midlands as a pork pig, it has been widely exported.

Duroc

A long-bodied, early maturing bacon pig of medium size, developed in the eastern USA during the 19th century from red-coloured European pigs introduced by Columbus. The red varies from golden to mahogany.

Hampshire

A long-faced, prick-eared pig of medium size introduced from the eastern USA and used mainly in cross-breeding. It is hardy and produces lean bacon and pork.

Piétrain

A very lean, dark-spotted pig bred for pork in Belgium during this century. It develops huge hams (thighs and buttocks), heavy shoulders and broad loins.

269

Short white hair

A high-yielding dairy goat with short legs and short white hair, sometimes with fringes of long hair. Some animals of both sexes are horned, but others have no horns. Medium, female about 121 lb (55 kg), male 165 lb (75 kg).

Short legs

The profile is dished (slightly concave) and the ears are erect with a slight forward tilt. Some animals have beards, and may have neck tassels near the throat.

Horned Saanen goats are not often seen. Most horned animals have the horn buds removed at an early age.

A female may have more than one kid at a time. The kids are usually hand-reared once they are four days old, and the female is then milked.

Saanen goat *Capra* (domestic)

Because of its ability to yield large quantities of milk, the Saanen is probably the most popular breed of goat throughout the world. Saanen goats originated in the Saane and Simme valleys in Switzerland, and were first brought to Britain in the late 19th century, but not on any large scale until the 1920s.

A Saanen female will often yield up to 330 gallons (1,500 litres) of milk in one year, and some have produced more than 440 gallons (2,000 litres) a year. Saanens will continue to give milk for a long time after giving birth, sometimes for two years. Goat's milk has a softer curd than cow's milk, and the fat rises more slowly to form a cream layer. It is more easily digested than cow's milk, so is especially valuable for feeding babies or for people with digestive problems.

Saanen goats are sometimes seen tethered by the roadside, for they are often kept by householders for a private milk supply as well as in commercial dairy herds. Goats are natural browsers rather than grazers, preferring shrubs to grass. Whether kept by the roadside or in a garden, they may need to be tethered not only to stop them from wandering away but also to prevent them demolishing too many trees and bushes.

Drooping ears

Dip in back

Long legs

A tall, long-legged goat with high withers and hips giving it a characteristic dip in its back. The hair is very short and varies in colour. The ears are long and drooping and the tail is short. Medium, female 121 lb (55 kg), male 165 lb (75 kg).

These Anglo-Nubian goats have brown coats, which are unusual. Like all goats, they prefer shrubs to grass.

Anglo-Nubian goat *Capra* (domestic)

Nubian goats from North Africa were introduced to Europe in 1860 when the King of Abyssinia presented a young hippopotamus to Napoleon III of France. The goats were used to supply milk for its journey. Nubians, used to the heat of Africa, could not tolerate the colder climate of northern Europe, and in Britain they were crossed with native goats to produce the Anglo-Nubian – so-named in 1890 – the oldest recognised British breed. Later, Zaraiby goats from Egypt and Jumna Pari goats from India were also interbred to produce the modern Anglo-Nubian, which is noted for the richness of its milk.

Its long neck and lop-ears make the Anglo-Nubian look different from any of the other British breeds, which are mainly of Swiss origin. The colour of the coat varies from white through various shades and mixtures of tan and brown to black. The commonest colours are tan, reddish or black, each of which may be mixed with white. The Anglo-Nubian's long-legged, long-necked build is considered ungainly by some people. Some animals also have defects such as a twisted jaw or dropping udder, but these are being improved by selective breeding. Some are horned, but breeders prefer hornless animals.

Although the milk yield is low compared with some breeds, the milk is high in butterfat and proteins. A female can produce about 275 gallons (1,250 litres) of milk during a year.

An Anglo-Nubian goat has a strongly convex profile with a prominent forehead and a high-bridged nose. It lacks the neck tassels that are found on some other breeds.

Goats for milk and yarn

Modern breeds of goat in Britain are either foreign breeds that were imported in the late 19th or 20th centuries, or breeds derived from crossing the imported breeds with native goats. Characteristic native goats were almost lost through such cross-breeding, but have been re-established as the English breed. Despite attempts to breed out horns, many goats of both sexes are born with horn buds, but these are normally removed a few days after birth. In several breeds, goats have a tassel of hair at each side of the throat. Many also have a beard on the chin – both females (nanny goats) and males (billy goats). Most breeds in Britain today, including the Saanen and Anglo-Nubian (pp. 270–1), are kept for milk and cheese, usually on a small scale in groups of less than six.

Bagot
Crusaders brought this horned, hardy breed, probably of Swiss origin, to Britain in the 13th century. Until the 1970s it was kept in semi-wild groups by the Bagot family of Staffordshire. The amount of black on the front parts varies.

Toggenburg
A small, fawn goat with distinctive white patterning. It was brought to Britain in 1884 and was the first Swiss breed to be imported.

Golden Guernsey
Syrian or Maltese ancestors may have given this rare, heavily built milker its deep orange-red skin. The ears are distinctive, pointing sideways and then turning up at the tips.

British Saanen
Bred in Britain from the Saanen and native goats, but larger than the Saanen and with bigger ears and a straighter face profile. The milk yield is higher than in any other breed.

Angora

A native of Turkey, the Angora is now kept mainly in Australasia, South Africa and North America. It is valued for its silky coat with fibres up to 8 in. (20 cm) long, which are spun into mohair yarn.

English

A long-bodied, horned breed with a tapering head and a dished (concave) face profile. Its coat is often fawn with a dark back-stripe and dark marking on the legs, but it may be grey, black or white.

British Alpine

Common ancestry with the Toggenburg has given this breed the same colour pattern but with a short, black, glossy coat. Some animals have a light-coloured belly.

British Toggenburg

Usually a much larger goat than the Toggenburg, with a longer, straighter face, bigger ears and a shorter coat that is often also darker. In Britain it gives a better milk yield than the Toggenburg.

Many common farmyard fowl are of the Old English Game type, kept as much for interest as eggs.

The hen's narrow and fan-shaped tail is carried upright. Long quills edge the large and powerful wings.

Small, fine wattles (or skin flaps) hang at the hen's throat. Her single small comb is upright. A cock's comb and wattles are sometimes removed.

Long, curved tail

Low-set spurs

The black-breasted red is popular among the colour varieties of Old English Game fowl bred today. The cock's tail is long, curved and spreading and its legs are hard and muscular with hard spurs set low. It has three toes in front and one behind. Small, cock 6 lb (2·7 kg), hen 5 lb (2·3 kg).

As with many poultry breeds, Old English Game fowl have a miniature, or bantam strain, often kept for showing.

Old English Game fowl *Gallus* (Domestic)

Fowl have been kept as sporting birds in these islands for about 2,000 years. In the 1st century AD the Roman conquerors recorded that the British kept native fowl for pleasure and diversion. They were in fact kept for cock-fighting, which remained a popular pastime for centuries and continued even after it was declared illegal in 1849. The fighting fowl, known later as the Old English Game breed, was one of the chief ancestors of British breeds of fowl. Landowners who bred gamefowl often boarded out young cocks with local tenants and cottagers. As a result many barnyard fowl came to be of the Old English Game type and continued to be so, especially in northern England, until the controlled breeding of recent times.

Today the breed is kept mainly for exhibition and is bred to standards laid down by the poultry club; 24 main colour varieties are recognised. A bold and alert bird with rapid, graceful movements, the Old English Game fowl owes its aggressive nature, hawk-like beak, firm muscles and hard, glossy plumage to its fighting ancestors. Even today cocks may have their comb and wattles removed, or dubbed; this was done for cock-fighting because it denied an easy hold to the bird's opponent.

There is a light-coloured 'bean' at the end of the bright orange beak, which joins the head smoothly with no knob. The eyes are light blue.

Plump wide breast

A large white goose with a long body and broad paunch, the Embden carries its head and plump, wide breast high. It has long wings that cross over on its back, in front of its tail, and short, bright orange legs. Large, gander 30 lb (13·6 kg), goose 26 lb (11·8 kg).

Short orange legs

Like all geese, the Embden needs enough water to swim, not only to keep its plumage clean but to aid fertility. The females are diligent mothers.

A paddock or orchard provides the geese with the ample grazing space they need, but it must be well fenced.

Embden goose *Anser* (domestic)

The hardy Embden goose will thrive with a minimum of attention and shelter but has a hearty appetite and needs plenty of space to graze and forage. It is an aggressive breed and will not mix easily with more placid geese. Nor is it easy to keep confined, for although it is large it is active and agile and will fly more readily than other geese. Embden geese are wary and alert, characteristics that make them effective sentries – quick to call the alarm at the slightest disturbance.

Although some of the Embden's ancestors came from Germany and northern Holland – their original locality is still remembered in the bird's name – the breed was developed mainly in England by cross-breeding with the English White. Noted chiefly for its meat, the Embden is a rapid-growing goose with a lean carcass, and does not lay as many eggs as some other heavy breeds, averaging about 25 a year. Embden ganders are commonly mated with geese of more prolific breeds in order to combine the best qualities of both. The Embden's pure white feathers and down are used to fill quilts and pillows. Goslings have pigment in their down, heavier in females than males, so sex identification is possible at hatching time.

Typical breeds of poultry in Britain

Poultry are kept for meat or eggs, some breeds for both. Apart from the Old English Game fowl and Embden goose (pp. 274–5), there are a number of typical farm breeds. British breeds of fowl have been developed from lightweight breeds brought from southern Europe or heavyweight breeds from the Far East. In recent years many breeds have been kept more for ornament and for shows than for use. Birds for intensive battery (eggs) and broiler (meat) units are usually hybrids.

Croad Langshan fowl

A heavy farmyard breed suitable for the table and also kept for its eggs, which are light brown and laid mainly in winter. Croad is the name of the first importer; the bird originated in Asia.

Toulouse goose

A heavy table breed reared for centuries in southern France, where its liver is used for pâté. It has a deep body that is carried horizontally, and a large dewlap below its bill. Females weigh up to 22 lb (10 kg).

Buff Orpington fowl

A white-legged, heavy-bodied bird with a white or yellowish beak. It is a good layer, especially in winter, and also acceptable as a table bird. The eggs are light brown.

White Leghorn fowl

A hardy lightweight laying breed with pure white feathers and a slender, erect body. Its eggs are white, and it lays most during summer.

Old English Pheasant Fowl

A lightweight active breed often seen in farmyards in northern England. It is kept for both eggs and meat. Until this century it was known by various names, such as Golden Pheasant and Yorkshire Pheasant.

Rhode Island Red fowl

A heavy table bird with a deep breast and a long back, introduced from the USA about 1909. It is also a good winter layer. The eggs are light brown.

Scots Grey

A hardy, lightweight breed of Scottish farmyards. It is a very heavy layer, especially in spring and summer. The eggs are white.

Chinese goose

A small, plump table bird with an upright body. It has a prominent knob where the bill meets the head. There is also a white variety with an orange bill.

Hen

Light Sussex fowl
A fine-textured, full-flavoured table bird with a broad breast. Its colouring varies, but is commonly white with black on tail, wings and head. It lays well in winter; the eggs are cream.

Cock

Hen

Marans fowl
A French-bred fowl that lays large brown eggs, mainly in winter, and is also of medium quality as a table bird. It does well scratching in field and farmyard.

Cock

Bronze turkey
Bred in Britain from native North American turkeys, the bird has a distinctive greeny-brown plumage. Its legs are either black or buff. Cocks weigh up to 40 lb (18 kg).

Duck

Aylesbury duck
A prime table bird that weighs up to 9 lb (4 kg). It is full-breasted, quick-growing and pale-fleshed with a broad, long body that is held horizontally.

Drake

Muscovy duck
A plump-breasted table bird descended from the South American musk duck. It has a very long, deep body which is held horizontally. Its colouring may be blue, white, black or mixed, or black with white-barred wings.

Duck

Indian Runner duck
Introduced from Malaya in the 19th century, the lightweight duck is kept for its eggs. It holds its long, narrow body erect. Its colouring may be white, black, brown, fawn or mixed.

Cock

Indian Game fowl
A table breed with ample breast meat. It has short, close-lying plumage and short thick legs. It was developed mainly in the West Country and in the 19th century was a much taller bird.

Cock

Dorking fowl
A heavy, quick-growing table breed with a very deep and well-fleshed body. It is unusual in having five toes instead of four. There are also red and light grey types.

White collar

White face-stripe

A loyal, intelligent working dog, used to herd sheep. Usually a Border Collie is black with white markings, but some dogs have tan patches as well. White markings are commonly on the nose, face, collar, chest, front legs and tail-tip. About 22 in. (55 cm) high at shoulder.

A dog's speed and skill in driving a flock or gathering strays makes it indispensable for sheep herding.

Some dogs are long-haired. All dogs usually have a broad, square head and semi-erect ears with the points turned over. But the ears are often erect when the dog is alert and waiting for an order.

A dog controls sheep partly by its 'strong eye'. Even a ewe defending her lamb will slowly give way if a dog fixes her with a commanding stare.

Border Collie *Canis* (domestic)

Men first trained domesticated dogs to help them hunt wild animals in the Stone Age. Later, early farmers used dogs to help with guarding and herding other domesticated animals. For the dog, hunting and herding are closely related. Just as a hunting dog instinctively drives the quarry towards the pack leader, so a herding dog drives sheep or cattle towards its master. This ability is most highly developed in the Border Collie, one of the most used sheep-dogs, and it is strongly inherited. Some Border Collie puppies have such a powerful herding instinct that they will round up ducks, chicks or other small creatures.

Border Collies probably developed as a breed on the Scottish border from medieval times. They are agile and very intelligent, and can be trained to undertake quite complicated manoeuvres. A good dog can, on command, separate one particular sheep from a flock, or suppress its natural instinct and drive sheep or cattle away from its master. The selection of a good dog depends on its performance rather than its appearance, so Border Collies vary much more in colour, size and build than other breeds seen mainly in the show ring. They usually excel in working sheep-dog trials and obedience tests.

Large head

Black coat

Square muzzle

Landseer Newfoundlands are white with a dark head (except for a white blaze), a dark saddle-patch, and a dark rump-patch extending down the tail.

Strong, intelligent Newfoundland dogs are natural swimmers and can be trained to save human lives.

The massively built Newfoundland dog, once used to pull carts and boats, is docile despite its size. It is usually black and has a large, broad head with drooping ears and a short, square muzzle. Male 28 in. (70 cm) high at the shoulder, female smaller.

The dog's skin stays dry even after a long time in the water. This is because its strong, straight outer coat protects an undercoat of soft, dense hairs that trap an insulating layer of air against the skin.

The Newfoundland is often used for boat towing. The webbing between its toes helps it to propel itself through water.

Newfoundland dog *Canis* (domestic)

When Leif Eriksson led a Viking expedition to Newfoundland about AD 1000, a Norse Saga records, he took with him a large black bear-dog. This was probably the forerunner of the New-foundland breed. Settlers from Europe brought many different breeds of dog to North America from the end of the 15th century, and many of the larger ones probably played a part in the development of the Newfoundland. It was a working dog, used for haulage and – because of its swimming ability – to carry lines from fishing boats to the shore or to rescue men overboard. The dog as it is today belongs to the mastiff group of breeds and is kept mostly on farms and estates. The Newfoundland Club holds regular water-trials to test a dog's life-saving ability.

Newfoundlands first came to Britain in the mid-18th cen-tury, and because of their good temper were often used as children's companions. Until superseded by the Labrador, they were used as gun dogs. They also used to pull carts until 1850, when it was forbidden by law. Most Newfoundlands are black, but a few are brown. There is also a black-and-white or brown-and-white variety known as Landseer after Sir Edwin Landseer, the 19th-century artist who often painted them.

A landscape dependent upon sheep

Many upland landscapes – wide expanses of dull green grassland – have been created largely by the hardy hill sheep, such as Swaledales, that live there, thriving on grazing that is too poor for other domestic animals. Long ago many of these uplands were well wooded, but sheep eat and trample woody plants and centuries of grazing have prevented the growth of tree seedlings and limited the number of shrubs. Sheep hoofs treading on soft peat destroy the grass in places, and soil erosion begins.

Hillsides closely cropped by sheep have little food or shelter to offer small mammals such as voles. Big black slugs, which feed at night, are the main competitors for the hillside grass and can be so numerous that they eat as much of it as the sheep. Small mammals live in the longer vegetation by drystone walls, but sheep scrambling over the old walls break them down. Many farmers have replaced walls with cheaper wire fences that give no shelter to plants or animals.

Adult hill sheep have no predators in the wild, but badly trained pet dogs cause the deaths of hundreds every year by chasing and worrying them. Severe weather may kill others. A dead sheep provides welcome meals for foxes and crows, both of which also kill lambs. When a sheep dies, the farmer loses not only its produce but also the government subsidy paid for each hill sheep reared.

A well-trained border collie, working to calls or whistles from the shepherd, will not press the sheep too closely, or they might panic and run the wrong way.

Moles are common on upland pastures and their tunnelling helps to drain and aerate heavy soil. But molehills smother grass, and if numerous rob sheep of a large amount of grazing.

Trees are sparse where sheep graze because they nibble seedlings and saplings and prevent growth. Only in fenced-off woodlands can new trees grow to replace the old.

Sheep shelter behind walls, and their droppings enrich the soil. Nettles and other plants flourish there, giving cover to mice, voles and shrews. Hedgehogs find the short, sheep-grazed turf an easy foraging ground at dusk.

The herding instinct that keeps the sheep in a group as the collie approaches makes them easier to control.

Typical working dogs

Hunting and herding have for centuries been the chief tasks for which men have used dogs. In the past some have been used for other tasks; the Newfoundland (p. 279), for example, once pulled carts or boats. Dogs such as the Border collie (p. 278) are still widely used to herd sheep, and today many dogs are used as guard dogs, guides for the blind or to sniff out drugs or explosives. Companionship between man and dog grew through working together, and many breeds once valued as workers are now commoner as family pets.

Greyhound
Since the days of ancient Egypt, the greyhound has been used to catch small, running game such as hares, which it hunts by sight. It is now kept mainly for track-racing. Colours: fawn, grey, black, white, brindled, sometimes with white markings. About 29 in. (73 cm) high at shoulder.

Old English Sheepdog
The dog's dense, shaggy coat protected it in harsh weather as it herded or watched over sheep and cattle, or drove them to market. It has a distinctive gait because it moves both legs on the same side at once. Colour: white-patterned grey or brown. About 22 in. (55 cm) high at shoulder.

Irish wolfhound
The fast, strong ancestors of the Irish wolfhound were used to hunt wolves in Britain 2,000 years ago. Later they were kept for hare-coursing. Under its rough coat the dog is built like a greyhound. Colour: brindled, brownish-black. At least 31 in. (79 cm) high at shoulder.

Foxhound
For centuries, foxhounds have been used in packs to hunt foxes by scent, and have the speed and stamina to run all day long. They often bay in deep, hollow tones as they catch a scent. Colour: white, black and tan markings. About 23 in. (58 cm) high at shoulder.

Alsatian

First seen in Britain in 1911, alsatians were originally German shepherd dogs. Intelligent and devoted to one owner, they now serve mainly as police, guide, sniffer and guard dogs. Colour: varied. About 25 in. (64 cm) high at shoulder.

Fox terrier

Used to drive out a hunted fox or badger from its hole, a terrier is small enough to enter a den but strong enough to keep up with the hounds. Colour: white with brown or grey markings. About 15½ in. (39 cm) high at shoulder.

Golden retriever

The breed originated in the north of England, and has been used as a gun dog for 100 years. Retrievers are now often used as guide or sniffer dogs. Colour: sandy brown. About 23 in. (58 cm) high at shoulder.

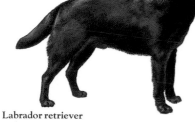

Labrador retriever

Its dense, water-repellent coat protected the dog in the icy Canadian seas, where it was used to retrieve fish that fell from nets. Labradors have been used as gun dogs in Britain for 150 years, and are now used also as guide or sniffer dogs. Colour: black, yellow, brown. About 22 in. (55 cm) high at shoulder.

Springer spaniel

The springer's task is to work through undergrowth to flush out game birds and drive them towards the guns. Tall for a spaniel, it runs well and is also used to retrieve dead birds. Colour: white with liver, black or tan markings. About 20 in. (50 cm) high at shoulder.

Bloodhound

The Normans brought this dog to Britain for stag-hunting. It has a well-developed ability to follow an animal trail by scent, and was once used to track down criminals. Colour: usually red-brown. About 26 in. (66 cm) high at shoulder.

Powerful
body

Thick
neck

The lower legs have a lot of
feathering; the feet are white.
There are competitions for the
best-turned-out heavy horses
at many agricultural shows.

Brewery Shire mares are sent to the country to foal. All dray horses have an annual
country holiday.

The Shire is the largest British breed of horse
and often weighs about a ton. It has a short,
wide powerful body and a thick, short neck.
Black, brown and bay are the commonest
colours, but grey, roan and chestnut occur.
Stallion 17 hands (68 in., 173 cm) high at
withers; mare slightly smaller.

On some farms, horses are
being used for ploughing
once again.

Many Shires are used by breweries; they are cheaper
than motor vehicles for hauling drays on short journeys.
When harnessed in pairs, each horse prefers to work
on a particular side.

Shire horse *Equus* (domestic)

The Shire's ancestors were war horses – the great horses that
carried armoured knights into battle. The breed was developed
mainly in the fen country of Lincolnshire and Cambridgeshire,
and gained its great stature from the massive Flemish stallions
that were imported in the 1200s and 1300s and mated with
English mares. As cannon displaced knights in armour, the
heavy war horse changed its role from mount to harness animal,
pulling loads such as artillery and supply wagons. It was not
until the 18th century that the Shire – more widely known then
as the English cart-horse – became a farm horse, replacing the ox
as a draught animal for pulling ploughs and hay wagons.

During the 20th century some Clydesdale blood was introduced to the breed to reduce the excessive feathering – hair on
the lower legs – that had developed among Shires bred for the
show ring and is a disadvantage in muddy conditions. The Shire
was superseded by the tractor during the Second World War, but
in recent years Shire numbers have increased, chiefly because of
their use as dray horses by breweries. Shires begin their working
life at four years old and give up to 16 years' service. The biggest
animals can pull loads of up to five tons.

Short legs

Heavy head

The hardy pony thrives on poor upland pasture. Its hind feet are often white, unlike those of the similar Fell pony.

Sturdily built and short-legged, the Dales pony is usually black, brown or bay, sometimes grey. The head is heavy and less refined than in most ponies, and may have a white star. The lower legs have long hair (feathering). About 14·2 hands (58 in., 147 cm) high at withers.

Some hill farmers still use the Dales pony for pulling loads in narrow lanes, as well as for working the land in fields of an awkward shape.

Dales ponies have powerful legs and are strong and sure-footed. They are often used for pony-trekking and for trap-driving competitions.

Dales pony *Equus* (domestic)

Today, Dales ponies carry pony-trekkers comfortably astride their broad backs along the tracks and lanes of the Yorkshire and Durham dales where for centuries their ancestors were pack and draught ponies. As the chief means of transport in the area, they pulled ploughs and loads such as hay, carried fleeces to the mill and fodder to snowbound sheep, and drew the farmer's trap.

The sturdy ponies are built like small-scale heavy horses but are cheaper to feed and nimbler in the steep, winding dales lanes. They once carried panniers of coal from the mines to the lime-kilns, and in the 18th and 19th centuries transported lead from the Swaledale and Nidderdale mines to Teesside to be smelted for the shipyards. Ponies trudged 250 miles (400 km) a week, each with a 440 lb (200 kg) load of lead. Their high leg action helps the animals to avoid the tussocks and ridges of moorland tracks, and the feathering on their legs throws off rain that trickles down their thick, slightly oily coats. Even working hard in all weathers they are long-lived, often reaching 30 years. Dales ponies are closely related to Fell and Highland ponies. In the 20th century some were cross-bred with Clydesdale horses to increase their size, and it also gave them longer feathering.

Typical working horses and ponies in Britain

For as long as men have herded flocks and tilled the soil, horses and ponies have worked for them, either as mounts or as pack or draught animals. There are several typical breeds of pony other than the Dales (p. 285), Exmoor and New Forest (p. 75) and Dartmoor (p.289), and a number of heavy horses apart from the Shire (p. 284). Height to the withers is given in hands: 1 hand equals 4 in. (10 cm). Ponies are never more than 14·2 hands (14 hands 2 in.), horses are larger.

Shetland pony
An extremely hardy and powerful saddle and pack pony, the Shetland grows a long winter coat with a woolly undercoat. Its colouring is varied. 10 hands.

Clydesdale horse
The long-legged Clydesdale is slimmer, faster and more agile than other heavy horses, but not as strong. It may be bay, brown or black, often with white on the face and legs, which are feathered with long hairs. 17 hands.

Eriskay pony
The Eriskay is similar in its colouring and sturdy build to the Highland pony. But in the harsher, more exposed island conditions, it has developed as a smaller breed. 13·1 hands.

Caspian pony
Once used by Persian peasants to pull carts, the Caspian is descended from the same ancestors as the Arab horse. Swift, agile and fine-boned, it has narrow, oval hooves hard enough never to need shoes. Now rare in Iran (Persia), its main home is in Britain, where it is used as a riding pony. 11·2 hands.

Connemara pony
A short-legged, deep-bodied pony hardy enough to survive on poor mountain and shore vegetation in western Ireland. Grey is the commonest colour. 13·2 hands.

Welsh cob
A small, strong horse developed from the native ponies for draught work on farms. The cob's quick, nimble movements made it also suitable for carriage work. It is similar in colouring to the Welsh Mountain pony. 14·2 hands.

Welsh Mountain pony
Cross-breeding with Arab horses has refined the build of this ancient breed. It is sure-footed, alert and strong. Once used as a pit pony, it is now a saddle pony. Its colouring is often grey but may be black, cream or many shades of brown. 13 hands.

Cleveland Bay horse
Bred in Yorkshire from the sturdy horses used to pull coaches and carry merchants' goods, the Cleveland Bay was refined with thoroughbred blood as a specialist carriage horse. It is bay or brown with black feet and a black tail. 16 hands.

Spotted pony
Prehistoric cave paintings depicted spotted horses, but the colouring is now rare except among gipsies' animals. Most modern spotted horses and ponies are of Spanish origin. The spots feel raised on the coat and may be dark on a light coat or light on a dark one. 12 hands.

Suffolk horse
A very broad-built heavy horse with speed as well as pulling power, often called the Suffolk Punch. It is hardier and needs less food than other heavy breeds. It is chestnut, sometimes with a white star on the face, and has little leg hair (feathering). 16 hands.

Fell pony
The Fell pony of the western Pennines is very similar to the Dales pony of the eastern Pennines. It is a large, black or dark brown pony used for pulling farm machinery and other loads. 13·3 hands.

Highland pony
Deep-chested and powerful, the Highland pony is used for carrying deer carcasses, pulling timber and other heavy work. It is often dun with a dark stripe along the back, but may be black, brown, grey or chestnut. 14·2 hands.

Hardy ponies of the upland moors

Small and sturdy Dartmoor ponies make excellent riding ponies. Many run wild on their native Dartmoor and are rounded up yearly in autumn by their owners for branding. Suitable ones are kept for selling, others are returned to the moor for breeding. The ponies are hardy enough to withstand the worst of Dartmoor's winter weather, scratching through the snow to find food when necessary. Cattle grids on access roads confine the animals to the moor.

Because there is no restriction on the breed of stallion that can be turned out on the moor, few of the hardy moorland ponies are pure-bred. At one time Shetland stallions were introduced by miners in order to breed pit ponies. Today most pure-bred Dartmoor ponies are kept in private studs and are no longer hardy enough to winter on the moor. Standards for the size, colour and conformation of the breed are set by the Dartmoor Pony Society. Piebald (black and white) and skewbald (brown and white) ponies are not acceptable, although some animals have small white markings on the head and legs. The maximum height at the withers is 12·2 hands (50 in., 127 cm).

The stallion is ever watchful for danger, rivals or straying mares. Stragglers are chased and soundly nipped. When his own filly (female) foals are about two years old, they are driven off to join another herd.

Its strong legs, tough feet and sturdy body enable the Dartmoor pony to survive on the moor's rock-strewn slopes and amid its numerous bogs. When foraging for bog grasses, ponies may sink belly deep into the soft ground.

Breeding herds of Dartmoor ponies run free on the moor. They are led by a stallion who keeps together a number of mares and foals, a herd ranging from about five up to 30. The ponies are long-lived, having a normal life-span of about 25 years.

Ponies spend most of their time grazing, each herd within its own territory, and will travel many miles in search of food. As well as moor grasses, sedges and rushes, the ponies will eat heather and even gorse if no other food is available.

Starlings are often seen among a herd of ponies. The flies that gather round the ponies provide the birds with a rich food supply.

Most Dartmoor ponies are bay, brown or black. A few are grey or chestnut. The thick mane and long tail help to protect them from rain, and in August they start to grow a thicker coat that remains until May.

WATCHING
AND STUDYING
ANIMALS

How to see and study animals

The best way to learn about wild animals is to watch them in their natural surroundings. To do this with any success, you need careful preparation and some basic skills, because relatively few species are active in daylight and fewer still move about slowly enough to allow observation for more than a few moments. They are also very shy, and will usually detect you before you see them.

What to wear

Although the colour vision of many animals is limited, they are quick to notice differences in tone. Brightly coloured clothing may be very conspicuous to an animal, so do not set out to look for animals wearing a bright red anorak, for example. Generally it is best to wear dark green and brown, and avoid sharply contrasting tones.

Some camouflage jackets can be effective, but many are too brightly patterned. Avoid synthetic waterproofs which rustle as you walk, and be sure that what you wear is warm and comfortable. Watching at a site may involve sitting still for hours; cold, uncomfortable clothing makes this difficult.

Using binoculars

A good pair of binoculars is essential equipment for watching animals. Generally, powerful telescopes are not very useful, unless the animals stay very still; binoculars are better. Different types vary in power and light-gathering ability, features that are described by a pair of numbers stamped on them, such as 8 × 30 or 10 × 50. The first number of the pair indicates the magnification. Less than 6 is insufficiently powerful; more than 10 is too powerful and it will be difficult to use the binoculars without a tripod to steady them.

The second number should be as large as possible because it is a measure of the size of the front lenses (in millimetres) and their light-gathering ability; 8 × 30 and 10 × 50 are best for nature-watching. Normally 8 × 20 will not give good performance in poor light, except with the modern miniature binoculars designed to slip into the pocket; these use a sophisticated optical design to achieve a bright image. They are very light and convenient, but are much more expensive than conventional binoculars of comparable quality.

Stalking and waiting

Most people move too noisily and too fast to see wild animals, let alone study their behaviour. Learn to stalk animals by walking very slowly, stopping every so often to listen and look about you. Be careful to make as little noise as possible – avoid walking on dry leaves and brittle twigs. Two people will obviously make more noise than one, and there is a great temptation to whisper together. Resist it. Keep together and stay where there is cover, avoiding well-lit areas and skylines. If you see anything, stop and stand still.

Animals have a very keen sense of smell, and it is impossible to approach closely if the wind is blowing your scent in their direction. Test for wind direction by tossing crumbled dry leaves, or something similar, into the air. If it blows away from the animals it may be possible to get a little nearer without being detected.

When approaching animals across open ground, or if they have seen you already, do not look at them directly or walk straight towards them. Approach at an angle and stare at the ground as you walk slowly along, look away frequently and glance at them only briefly from the corner of your eye. Such ruses will often allow a closer approach than crawling or trying to hide behind trees. The animals will almost certainly see you doing that, and when they do they will be frightened away by your unusual behaviour. Do not wave your arms about or swing them vigorously, as this appears threatening. Keep them close to your sides.

Many of the most rewarding observations of animals have been made by simply sitting and waiting, either at known sites such as badger setts, or in areas where animals have been heard or glimpsed. It is often better to wait for animals to come close than to stalk too near and risk scaring them off.

Many dedicated naturalists use hides, made of brushwood, canvas or camouflage netting. Although very useful for long-term observation at well-protected sites, hides have many disadvantages. Their construction can be disruptive in itself, and if left up they retain human scent which would otherwise blow away on the wind. They also attract the attention of vandals. A hide erected beside a fox's earth, for example, may indirectly cause the destruction of the earth and its occupants. Where there is road access, a car makes a good temporary hide, and has none of the drawbacks of the permanent type. Animals such as deer and rabbits will often allow a slowly moving car to approach quite closely. But do not get out or make a noise.

Most animals (including humans) rarely look up, so sitting up in a tree above a badger track or overlooking a deer feeding place will often permit a close view of animals. The human scent drifts away high over their heads, so the direction of the wind is less important. Special 'high seats' equipped with ladders are used in nature reserves and forestry plantations, and are very effective.

Such refinements are not essential, however, for successful animal observation. Simply sitting or standing still and blending with the background is quite sufficient, assuming the wind direction is right. The most important asset is patience.

Seeing in the dark

A lot of animal activity goes on at night, and devices that help you to see in the dark can be very useful. Professional naturalists use equipment originally developed for police and military use, such as electronic image-intensifiers and binoculars which can 'see' a scene illuminated by infra-red light (which is invisible to animals). Such equipment is extremely expensive.

The amateur can improvise by using a torch fitted with a red filter made of transparent red plastic, or by coating a bulb with red ink or layers of dye from a red felt-tip pen. Humans can see reasonably well in red light but nocturnal animals cannot, and are not aware that they are being illuminated.

You can deceive an animal's eyes by using a red torch, but not its ears and nose. It is still necessary to keep quiet and avoid allowing your scent to drift on the wind towards the animal.

Some hints on photography

Photographing wild animals is a challenge even to experts, and can involve a lot of expense and frustration. Nevertheless, it can be a useful way of making a record, and the intention to take a picture is a good incentive to careful stalking and observation.

Large animals such as deer can be photographed successfully with a simple 'snapshot' camera, but pictures of shy or distant subjects can be obtained only by using a telephoto lens – one that allows you to take a close-up shot from long range. Because of this, a camera that takes a range of lenses is virtually essential. Automatic-exposure types are recommended. In the heat of the moment it is very easy to make mistakes, and automation helps the mind to concentrate on essentials. Power-winding attachments are also useful, because when animals appear there is often very little time to wind on the film between shots; they are, however, heavy, noisy and an extra expense.

A zoom lens (an adjustable telephoto lens) allows you to take a close-up shot and then swiftly and silently change to a general view of the animals in their surroundings. Telephoto shots are often disappointing, as few wild animals will stand still in good light; use a fast film to 'freeze' movement.

At dusk or at night, some sort of illumination is necessary, and an electronic flash is ideal. The duration of the flash is very short, and used in moderation will not upset the animals. Photography at night is best done by waiting at a selected site, because it is not easy to illuminate a subject some distance away with a flash fitted on the camera.

For photography at a selected site, a sturdy tripod is a great help. It not only takes the weight of the equipment and prevents camera shake, but allows the camera to be aimed at a particular point (such as a rabbit hole), focused before darkness falls and locked in position. When the rabbit appears, all you have to do is press the button.

Good results can be obtained by having two flash guns mounted on separate tripods on either side of the camera. If you make one more powerful by wrapping a single layer of white handkerchief round the other, a faint shadow will be cast by the animal, emphasising its shape. The background will be black, but for nocturnal animals this is appropriate.

Most mammals are nocturnal, and many can be persuaded to feed at a regular baiting point. Food may be put on the patio for hedgehogs, peanuts in the hedgerow for mice and syrup and sultanas for badgers on a badger route, if you know they are about. The camera and flash can be set up on tripods near by and triggered by remote shutter-release. This is preferable to photographing animals 'at the nest' (badgers by their setts, foxes by their earths, for example). It avoids the risk of seriously disturbing the animals. If they decide to turn away, no harm is done. Do not forget that you need a licence to photograph specially protected species such as bats, otters and sand lizards in their places of shelter (see p. 297).

Preserving footprints

Although direct observation is the most exciting aspect of natural history, there is much to be learned from the study of animal tracks, signs and remains.

The footprints of small mammals such as voles do not usually show up well in mud or snow, so we are often unaware of their presence. By putting out something in which footprints will be clearly visible, you can see if there are any about.

A sensitive surface for detecting mammal footprints can be made by taking a piece of flat glass or a shiny white glazed tile and depositing a thin layer of soot on it from a smoky candle. It should then be put on the ground (perhaps with bait near by) and covered with a tunnel made of bent tin, cardboard or half a plastic bottle, to keep the rain off and stop bigger animals walking on the footprint detector. Mice, voles and even tiny shrews will leave clear tracks in the soot.

The prints can be preserved by spraying them with hair lacquer. Better still, the marked sooty glass can be used as a negative for making a contact print on photographic paper. With the prints recorded, the glass can be cleaned, re-blacked and used again.

Often, people find burrow entrances and are not sure if they are in use. Prop a piece of twig loosely across the hole and see if it has been pushed aside the next day. If so, a larger piece of sooty glass, or a shallow tray of soft clay or mud, can be put at the entrance to record the prints of any animal going in. However, take care when interpreting the results. Dogs and foxes will often visit holes, push twigs aside and leave footprints, but they do not actually live there.

If you find good, clear footprints of larger animals, or can get them to walk on trays of soft clay, it is possible to make a plaster cast as a permanent record of the print. Bend a strip of cardboard into a ring and fix the ends together with staples or paper clips. Bed this into the soil around the print to enclose it in a wall about $1\frac{1}{2}$ in. (40 mm) high. Mix up some plaster of Paris and pour it into the footprint, prodding it gently with a stick to make sure that no air bubbles are trapped and the plaster trickles into every corner of the print. Fill the footprint and let the plaster spill over to fill the whole of the enclosed area right to the top of the card wall.

Let it set (it takes at least 10–15 minutes), then lift the circular slab of plaster off the soil, peel off the card strip and clean away any dirt with a soft brush. The shape of the footprint now stands out in relief from the slab of plaster, which can later be painted to make the print show up better. The species, date and place should be scratched on the smooth surface of the plaster.

Keeping skulls and bones

Cleaning skulls and bones from dead animals is a messy business which requires skill to be successful. On the other hand, skulls and bones which have been lying about for some time have usually been roughly cleaned by the activities of insects. Such bones can be cleansed of potentially harmful germs and whitened by treating them with well-diluted bleach. Household bleaches are a little harsh, and give the bone an unpleasant chalky surface, so it is better to use hydrogen peroxide (available from the chemist) diluted by ten times its own volume of water. Add some detergent for a good frothy bleach.

Bleaching time depends on size. A rabbit skull, for example, should be left in for about 12 hours. After bleaching, wash the skull and leave it to dry on a sheet of newspaper over a radiator for a day or so. Any loose teeth should be pulled out and then glued firmly back in their sockets before they get lost.

Making records

Keeping a diary or a notebook, with sketches and descriptions of what you see, is often fun and can make rewarding reading in later years. Dated records of species, with precise locations noted, can also be very valuable. No one knows whether or not a species is in need of protection unless its numbers and distribution are first recorded.

Nationwide surveys of species are organised periodically by the Mammal Society or, for reptiles and amphibians, the British Herpetological Society (see p. 314), and casual records of rarer species are welcome at any time. Detailed surveys are often undertaken by local societies or county naturalists' trusts, and everyone's help is needed.

An amateur can see, recognise and record a hedgehog just as well as an expert, so surveys are something everyone can join in. Distribution records are collated by the Biological Records Centre (Abbots Ripton, Huntingdon) to produce up-to-date distribution maps, or the more general distribution guides like those used in this book. Much of the information for such maps has come from the general public.

Finding out how far animals range

An animal's home range is defined as its normally used area, and includes places visited to feed or to search for a mate as well as one or more nest sites. Because it is difficult to measure or even observe exactly where an animal goes, it is not possible to set precise limits to its home range. This applies particularly to the nocturnal species, and to those smaller animals whose activities are either hidden by undergrowth or take place out of sight in burrows.

Similarly, an animal's territory (an area it defends) is not easy to measure. Mammals tend to advertise themselves and their territories to other animals by using scent markers, which are not obvious to humans.

Zoologists study home range and territory by capturing and marking animals so that they can be recognised, then releasing them. From records of marked animals recaptured away from the area where they were first caught, they can get an idea of where they go and how far the species can travel in a known period of time. Such studies also help to estimate the number of animals living in a particular area and, by implication, the population likely in one acre (or in one hectare).

Small mammals such as mice, shrews and voles are captured unharmed in specially designed traps, and can be marked by clipping patches of fur – ringing or tagging is not usually suitable. Clip marks are useful only for brief studies, however, because new fur grows at the next moult. Hedgehogs can be marked quite harmlessly by painting or clipping small patches of spines. Reptiles can be marked by clipping pieces off the belly scales; it does no harm and leaves a more or less permanent mark.

There are many drawbacks to studying small animals in this way. For example, if the traps are not baited, the animals may not go into them. If they are baited, the animals are provided with extra food so might not need to range as far as they normally would in order to get enough to eat. Also, the trap holds the animal for several hours before it is released, whereas normally it would travel about in that time. Such unnatural situations distort the study. Some animals such as dormice rarely go into traps, so cannot be studied by trapping and marking. Large animals such as deer are often marked with brightly coloured and numbered ear tags, so that recognition is possible from a distance. Numbered tags may also be attached to the hind flippers of seals.

It is now possible to fit an animal with a miniature radio transmitter (built into a collar) so that it can be tracked continuously and located at any time of the day or night without being disturbed. Such a transmitter can be smaller than a pea, but those attached to deer, for example, are the size of a matchbox and have a range of half a mile (1 km) or more. Smaller transmitters that can be fitted to lizards, frogs, toads, mice and even bats are becoming available. They will make it possible to learn a lot more about where and how far these animals range.

There are now also transmitters that can monitor body temperature, heartbeat and levels of activity as well as location. Detailed information like this is essential to a thorough understanding of animals, and will assist the formulation of proper conservation measures for rarer species.

Conserving and protecting wildlife

Britain's wildlife is under threat, not so much from hunters and collectors, whose actions are strictly controlled by law, but mostly from the loss of shelter and living space, from poisonous insecticides and factory effluents, and from litter such as cans, that may trap or wound animals.

Animal habitats are constantly being destroyed to make way for agricultural and building developments. For example, 35 per cent of deciduous woodland has been destroyed since 1945, up to 2,500 miles of hedgerows have been removed each year, and 10 per cent of farm ponds have been filled in annually in some counties. This not only robs many small mammals and amphibians of living space, but depletes the supply of insects and plants they depend on for food.

To help animals to survive, national and local nature reserves and forest reserves have been established, and some species, such as the sand lizard, are rare outside such reserves. The protection of caves and old mines where bats hibernate offers secure places for threatened species to live, as does the establishment of otter havens.

Other action that aids survival includes the provision of subways for badgers under roads and of escape ramps for hedgehogs and other small mammals in cattle grids. Also the posting of warning signs for motorists on roads where frogs and toads cross in numbers to reach breeding ponds in spring.

Rare breeds of domesticated animal are also being protected by conservation programmes. Many farms, parks and zoos (see p. 308) keep small groups of ancient or minority breeds not now reared commercially because they are not as profitable as specially developed modern breeds.

Rare breeds have valuable characteristics – size, milk quality, immunity to certain diseases, for example – which may be useful in future cross-breeding schemes when needs may well be different from those of today.

How you can help animals

Apart from supporting or helping some of the organisations working for wildlife conservation (see pp. 314–15), there are a number of things you can do to help animals.
Do not litter the countryside. Every year thousands of animals die trapped in bottles or cans carelessly thrown away. Clear up bottles and cans, and pick up plastic bags and cartons, which may be swallowed by deer or ponies and cause them to choke, or may suffocate animals that investigate them or try to nuzzle food from them.
Leave somewhere for animals to live. Paradoxically, too much tidiness can be as harmful to animals as carelessness with litter, because clearing out garden corners, grubbing up shrubs and mowing roadside verges destroys the living space of many small creatures. Try to tolerate selective untidiness. A pile of leaves left in a sheltered corner may provide somewhere for a hedgehog to hibernate. A piece of corrugated iron or similar material left in a grassy spot may become a nesting place for voles and wood mice. In a sunny place it enables reptiles to bask underneath, hidden from predators.
Avoid the excessive use of garden chemicals. Insecticides and slug pellets used too liberally may kill far more creatures than those that are damaging flowers or crops. Many small mammals and amphibians (birds, too) rely on insects as a food supply. The accumulation of poison from contaminated insects can lead to the death or decline of creatures that eat them, such as hedgehogs or bats.
Create habitats for various animals. Several species can be helped by building them a suitable habitat, and this also offers you a chance to study them. A grassy, undisturbed corner of the garden where the vegetation is left untouched will encourage butterflies and perhaps small mammals as well.

A garden pond will often attract frogs, toads and newts, and they will help to keep down insect pests. Bird nest-boxes will sometimes be used by squirrels and dormice. At least seven species of bat will use bat boxes; a bat box (right) has an entrance in the base. It can be fitted to a tree or wall.

If the box is fixed about 5 ft (1·5 m) or so above the ground it may attract long-eared bats or pipistrelles. At a height of about 16 ft (5 m) or more above ground, it may be used by large, high-flying bats such as noctules. Make sure a tree box is clear of branches. Generally, south-facing boxes are most likely to be used in spring and summer, and north-facing boxes in autumn and winter.

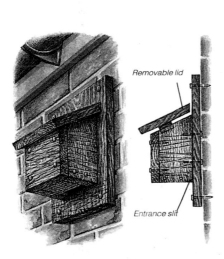

Removable lid

Entrance slit

A home-made bat box with an internal width and breadth of about 4 in. (10 cm) and a height of 8–12 in. (20–30 cm) can be made from roughly sawn timber about 1 in. (25 mm) thick. The ½ in. (13 mm) wide entrance slit should be in the bottom, extending the width of the box. Make a lid that slopes, to throw off rain, and that can be removed, so you can look inside. Do not coat the box with wood preservative; it may poison the bats. Horizontal saw cuts on the inside and outside of the box, and on the landing area below the entrance slit, will give the bats a foothold.

How the law protects wildlife

To keep Britain's wild animals free from being harmed or disturbed by people, most are protected by law. Species that are endangered and declining in numbers are given full legal protection. They are:

Otter	Bats (all species)
Red squirrel	Smooth snake
Wild cat	Sand lizard
Dormouse	Great crested newt
Pine marten	Natterjack toad

This means that it is against the law to:
Kill, injure, sell, capture or keep them (even for a short time for marking).
Disturb them in their breeding sites and places of shelter. Disturbing does not refer just to touching or moving the animals, but also to photographing them or doing anything that alarms them or rouses them from hibernation.
Destroy or damage their breeding sites or places of shelter, or block the ways in and out of them.
Although badgers are not rare or endangered, the same protection applies to them, except where there is licensed control of numbers. It is also against the law to ill-treat badgers, hedgehogs and other wild mammals.

The accidental killing or injury of a protected animal is not a breach of the law. If you find an injured animal of a protected species, you are allowed to keep it and take care of it until it can fend for itself again. If it is too badly hurt to recover, it may be humanely killed without breaking the law.

Until 1988 seals were fully protected only during the breeding season. For common seals this is during the whole of June, July and August, and for grey seals during the whole of September, October, November and December. At no time of year could seals be killed without a licence. Following a lethal virus plague that began in 1988, full legal protection was extended to cover seals all the year round, to help their populations to recover.

Animals partly protected

It is not against the law to kill or capture animals that are not fully protected, but there are regulations regarding the methods that may be used. Certain animals – the polecat, hedgehog, common, pygmy and water shrews – may not be caught or killed in traps or snares. Nor may they be killed or captured by the use of poisons, gas, lights, sound recordings or electrical devices. Also, water vole burrows must not be disturbed.

Certain cruel or indiscriminate methods are forbidden for the capture or killing of any wild animal. They include the use of self-locking snares, explosives (apart from firearm ammunition) and live decoys. Weapons such as longbows and crossbows are also forbidden. Permitted types of snare must be inspected at least once a day to see if any animal has been caught.

Among other species given partial protection, are the grass snake, the common lizard and the slow-worm. No native species of amphibian or reptile may be offered for sale, except with a special licence. This includes the adder, although it is otherwise less fully protected than the other reptiles.

Animals given seasonal protection

Deer are seasonally protected, generally during the summer months, but close seasons vary with species. Females are usually protected for a longer period of the year than males. The close season dates in Scotland differ somewhat from those in England and Wales. Culling (selective killing) is necessary to keep numbers under control, but is governed by strict regulations.

Dealing with pests

The law is not intended to prevent you from ridding your home or land of animals that are doing damage. You can set traps for rats and mice in your house or garden, but should take reasonable precautions to avoid killing pet animals or fully or partly protected wild creatures such as shrews, hedgehogs or birds.

Rabbits can also be snared, but snares must be checked daily. You may destroy foxes, grey squirrels and other non-protected animals if they are damaging your property, trees or crops, but only by humane methods.

Poisons can be used for rats, mice, moles, grey squirrels or similar pests around buildings, but there are strict controls on their use. Guns can be used only with a firearms licence, and again, there are strict regulations concerning them.

Dealing with bats in the house

Many of Britain's bats roost in buildings, often in modern houses, so they are the animals most likely to be disturbed by humans. It is against the law, however, to get rid of bats from the non-living areas of your house – attics or outbuildings, for instance.

Bats will not damage your property. They do not gnaw wood or bring in nesting materials, they merely hang, head down, from the roof or rafters. A bat's dry, powdery droppings, which consist mainly of insect fragments, are not a danger to health and make an excellent soil fertiliser. A single bat may eat more that 3,000 insects on a summer's night, including crop pests and wood-boring insects.

Most attics are too warm for bat hibernation, and are occupied by bats only at breeding time during June, July and August. If you consider them a nuisance, you must consult English Nature, Scottish Natural Heritage or the Countryside Council for Wales (see pp. 314-15) before making any attempts to block up the gaps through which bats enter roof spaces, eaves, the space behind hanging tiles, garages, sheds or other roosts. Nor must you use wood preservatives or other chemicals in the roosting places without advice from one of the three official conservation agencies.

Occasionally a bat may enter the living areas of your house. Even there you may not kill or harm it. Usually it will find its own way out if the window or door is left wide open. If a bat seems unable to find its way out or appears too drowsy to fly, lift it gently and put it outside on a wall or tree. Be sure to put on heavy gloves before you handle it; a frightened bat may bite.

Bats may be found hibernating in a cool cellar or outhouse during winter. Avoid arousing a hibernating bat, because it may die if it leaves the roost in winter.

Releasing alien animals

Non-native species are unwelcome in our countryside because they may rapidly take over the living areas and food supply needed by native species, which could become extinct as a result.

Some alien species have caused severe damage to trees and crops – the coypu for example. It is illegal to release into the wild any animal not normally resident in Britain, or to release any of the non-native species of mammals, reptiles and amphibians that have already established themselves in the wild here (see p.301).

Penalties and exemptions

Anyone who intentionally breaks the law can be fined up to £1,000, and if more than one animal is involved, each offence can be dealt with separately.

Licences may be granted for exemptions from particular parts of the law; for example, in order to photograph animals or capture them to ring or mark them for scientific or educational purposes. English Nature, Scottish Natural Heritage and the Countryside Council for Wales are the bodies responsible for administering the Wildlife and Countryside Act 1981, and can issue such licences. Where exemptions are necessary, licences may also be granted by Government agricultural departments.

How animals came to Britain

Until some 9,000 years ago, when the ice caps melted and the sea levels rose, Great Britain and Ireland were not the series of islands we know today, but were attached to the land mass of Europe.

With no sea as a barrier, it was possible for land animals to move freely across Europe and spread into the areas that are now England, Wales, Ireland and Scotland. Those that arrived in this way, before Britain and Ireland were cut off from the rest of Europe, are our native species. Seals and bats, animals not confined to land, are regarded as native if they breed in Britain.

Changes in climate

Before Britain became an island, the spread of animals both northwards and southwards across Europe occurred several times, and each time changes in climate forced the animals to move away again, or drove them into extinction.

In the last million years there have been at least four Ice Ages, during which only animals adapted to living in extreme cold, such as the now-extinct woolly mammoth, could survive. When the ice melted and the climate was warmer, the species that were adapted to cold moved farther north to be replaced by sun-loving animals from the south, such as the hyena.

The bones of some of these animals have been found in gravel beds and also in caves such as Kent's Cavern near Torquay, Devon. Caves in the Mendips have also yielded the bones of a pony that was probably the ancestor of the Exmoor pony; it had reached Britain before the last Ice Age.

About 10,000 years ago the last Ice Age ended, and with the thaw the animals gradually spread northwards again across Europe. They continued to do so as forests began to cover the land – first birch, pine and juniper and later oak and alder as the land became warmer and wetter.

The first animals to arrive would have been those able to withstand cold, such as the fox, otter, stoat, mountain hare, badger and red deer. Animals such as bats, hedgehogs, brown hares, weasels and common frogs would have followed as the climate became warmer. Some of the last native species to arrive would have been the mole, the red squirrel and the warmth-loving snakes and lizards.

By about 7000 BC, the melting of the polar ice cap had liberated enough water to raise the sea level and flood the swampy land between England and France and the low countries of the Netherlands and Belgium. Britain had become an island, and the only way land animals could now reach it was to be brought in by man.

Several animal species, such as the tree frog, salamander and garden dormouse reached northern France only after Britain had been cut off by the sea. Consequently they are found in Europe, only 22 miles (35 km) across the English Channel, but never became native to Britain. Although there are a few small colonies of tree frogs in England, they are recent introductions.

Animals introduced by man

A number of our more familiar animals were introduced to Britain either deliberately for farming, hunting or interest, or accidentally in cargo ships. Foremost among the useful species deliberately introduced are domesticated animals. The only ones that may have been domesticated directly from native British animals are some of the half-wild ponies – most likely the Exmoor pony – and breeds of pig such as the Welsh and British Lop, which are probably descended from the native wild boar.

Most domesticated animals – cattle, dogs, sheep, goats and poultry – probably arrived with Stone Age farmers at least 3,000 years ago. They were primitive types, and since then a great variety of different breeds has been developed. A few primitive breeds of sheep, such as the Soay, have survived on isolated islands. Cats, which originated in Africa, were probably brought to Britain by the Romans. Turkeys came from North America in 1521.

Although rabbits are now a familiar wild species, they were brought to Britain by the Normans in the 12th century because they were a useful source of meat and skins. In later centuries many country estates had their own rabbit warrens – land set aside for rearing rabbits. Not until the 19th century did rabbits come to be considered a pest, after crop-growing became more intensive and rearing them became uneconomic because of meat and skins imported from Australia. Domestic ferrets, used to drive rabbits from their burrows, are known to have been in use in Britain in the 1200s.

Fallow deer were brought to Britain for the chase and for venison, probably by the Romans. Roe deer, which were widespread in medieval times but gradually disappeared, were reintroduced about 120 years ago, probably for hunting.

One of the most recent reintroductions is the reindeer, brought to Scotland's Cairngorms in 1952. Reindeer were originally native animals and no one is sure when they died out in Britain. References to 12th-century reindeer hunts in the *Orkneyinga Saga* (a history of the earls of Orkney written about 1200) are questionable as they may have been referring to red deer.

Other species introduced for interest include the muntjac, Chinese water deer, grey squirrel and fat dormouse. Of these, the grey squirrel from North America is undoubtedly the most successful so far. It was deliberately released into the wild in the late 19th century, and has largely replaced the native red squirrel; in many forestry plantations it is controlled as a pest. A few muntjac and Chinese water deer may have been released deliberately from parks, but most of those living in the wild are descended from animals that have escaped from parks or zoos.

Marsh frogs were brought to Britain from Europe for use in laboratories, but some were released in a Kentish garden in 1935. Now they are widespread in the area of Romney Marsh and have been taken from there to various other places. Small colonies of tree frogs and midwife toads exist in one or two places. The tree frogs were introduced deliberately, the midwife toads by accident among imported water plants; neither species is likely to spread far because the British climate is unsuitable.

During the 20th century a number of animals were brought in to be bred for their fur, but some escaped or were released and have established wild colonies. The most successful has been the mink, which came from North America in the late 1920s and is now well established here. The coypu from South America escaped into the wild in the 1930s and became a nuisance because of its damage to crops; it is now extinct in its last home, East Anglia. The musk rat from North America, brought in during the 1920s, also damaged crops but had been exterminated by 1937.

Some of the most obnoxious pests – the house mouse, black rat and brown rat – were brought to this country by accident. The house mouse is probably the oldest inhabitant of the three. It is known to have lived here in the Iron Age, before the Romans, and may have been brought from Asia and the Mediterranean in cargo ships.

The black rat originated in South-east Asia, where it is known to carry bubonic plague through the fleas that live on it. The rat came to Britain on trading ships, probably in the 11th century. It has now largely been replaced by the larger and more adaptable brown rat, which reached Britain from Russia, also in ships, probably in the early 18th century.

The animals of Ireland

Even before Britain was cut off from the mainland of Europe, the Irish Sea had cut off Ireland from Great Britain. As a result, there are fewer types of land animal found in Ireland than are present in Britain, and probably only eight or nine mammals are native.

The explanation of how animals did or did not get to Ireland is still being debated by zoologists. Animals better equipped to withstand cold most likely reached Ireland before it became surrounded by sea. They are the pine marten, stoat, mountain hare, otter, fox, badger, red deer, the common frog and probably the pygmy shrew. The hedgehog may be native to Ireland but this is not known for certain.

Animals that prefer a milder climate – the weasel, and common and water shrews – probably spread westwards later and could not cross the sea barrier, so never reached Ireland. This may also be why Ireland has no moles nor, until recently, any voles or snakes. Bank voles were first discovered in south-western Ireland in 1964, and are now spreading. Grass snakes have been seen in Ireland in recent years, but how they or the bank vole were introduced is unknown.

Irish stoats are smaller and darker than mainland stoats, and have a more irregular line between the dark upper coat and the white belly fur. They look rather like a cross between mainland stoats and weasels.

Animals of offshore islands

The Scottish islands have even fewer native species than Ireland. This is because they are separated from the mainland by such deep water that they were probably isolated even before the sea level rose to cut Britain off from Europe.

Otters could have swum to the islands and are probably native species. Animals such as sheep and red deer, however, are more likely to have been introduced deliberately in the distant past. But the presence of small mammals such as mice and shrews, and amphibians such as newts, must be due to accidental importation. It is likely they were transported in prehistoric times with loads of peat and thatching materials from the mainland.

Wood mice were already present in Orkney by the New Stone Age some 5,000 years ago, as was the Orkney vole. The vole had been brought from somewhere in the Mediterranean region. Skeletons of both wood mice and voles have been uncovered in archaeological digs at New Stone Age settlements in Orkney. Later, the Vikings distributed wood mice widely among the Scottish islands, carried among the provisions in longboats. In this way also, house mice were introduced from Scandinavia.

White-toothed shrews are found in certain of the Channel Islands and Isles of Scilly, but nowhere else in Britain. They are common in southern and eastern Europe, and it was from the Continent that they were carried to Britain's most southerly islands, probably among imported fodder for farm animals, but nobody knows when.

Many island populations have developed from a small number of animals, or even, perhaps, a single pregnant female washed ashore with wreckage from a ship.

Because of their isolation from mainland animals, and because of different conditions in their new surroundings, island animals often gradually evolve a slightly different form. They become specially adapted to the particular conditions found only on their island home.

With no other similar mammals present, there is less competition for food, and in the absence of predators – the adder and weasel for example – they could become larger and tamer without risk. This has occurred with the Skomer vole, a form of the bank vole that has evolved on the island of Skomer off the Welsh coast.

The Orkney vole is also bigger than mainland field voles, and many island races of wood mice vary somewhat from the mainland animals, although not enough to be considered a different species. The most striking example is the St Kilda mouse, found on a remote island of the Hebrides. It is nearly twice the size of mainland wood mice, and has longer, thicker and greyer fur. Despite the differences, however, the fact that it will breed with mainland wood mice shows that it is of the same species.

THE WILD ANIMALS OF BRITAIN

Native animals	REPTILES
MAMMALS	Adder
Badger	Common lizard
Bank vole	Grass snake
Bats	Sand lizard
Brown hare	Slow-worm
Common	Smooth snake
dormouse	
Common seal	**Animals introduced**
Common shrew	**long ago**
Exmoor pony	MAMMALS
Field vole	Black rat
Fox	Fallow deer
Grey seal	Feral goat
Harvest mouse	House mouse
Hedgehog	Rabbit
Mole	Soay sheep
Mountain hare	White-toothed shrew
Otter	
Pine marten	**Animals introduced**
Polecat	**within last 300 years**
Pygmy shrew	MAMMALS
Red deer	Brown rat
Red squirrel	Chinese water deer
Stoat	Fat dormouse
Water shrew	Grey squirrel
Water vole	Mink
Weasel	Muntjac
Wild cat	Red-necked wallaby
Wood mouse	Reindeer
Yellow-necked	Roe deer
mouse	Sika deer
AMPHIBIANS	AMPHIBIANS
Common frog	Edible frog
Common newt	Clawed toad
Common toad	Marsh frog
Great crested	Midwife toad
newt	Pool frog
Natterjack toad	Tree frog
Palmate newt	

Britain's lost beasts

Many animals that were once common in Britain and Ireland are no longer found here. Some died out in comparatively recent times. With the development of farming and the loss of wild country, there was simply no room for them to exist alongside the spreading human population.

Others were inhabitants of Britain in the long-distant past when the climate was either tropical or in the throes of glaciation. They fell victim to the changes in climate.

Animals of the recent past

Bears and wolves roamed free in Britain until several hundred years ago. Brown bears were common during Roman times, and many were exported for use in wild beast shows. But they had died out by the early Middle Ages, killed by farmers who were concerned for the safety of their flocks and herds. They still persist in parts of western Europe, although they are now rare.

Wolves survived in England until about 1550, and small numbers were still to be found in Scotland and Ireland in the 1700s. Place-names such as Wolf Hole Crag in the Forest of Bowland in Lancashire recall their former moorland and mountain haunts.

Other large carnivores once native to Britain were the glutton (or wolverine) and the lynx. They probably died out in Britain during the Stone Age, some 5,000 or more years ago, but still exist in North America and the Arctic, the lynx also in a few remote parts of Europe.

Wild boars were once widespread in Britain, but died out here in the 17th century. Animals of the open woodlands, they gradually disappeared as forests were felled, and they were also relentlessly hunted for food and sport. They are still common in heavily forested areas of continental Europe.

Britain's beavers were once prized for their valuable fur and also their musk glands, which were believed to have medicinal properties. There has been no evidence of beavers in the wild in Britain since the 12th century, but they are recalled in place-names such as Beverley in Humberside. The beaver is still found in other parts of Europe, but is very rare.

The wild ox, or aurochs, was a very large ancestor of modern cattle and stood 6 ft (1·8 m) high at the shoulder. It was probably hunted to extinction in Britain during the Bronze Age, about 2,500 years ago, although in Europe it survived until the 17th century. The last wild ox died in Poland in 1627, but smaller animals that resemble it

Wolf *Canis lupus*

Wild boar *Sus scrofa*

Brown bear
Ursus arctos

Beaver *Castor fiber*

closely have been recreated at Munich Zoo, Germany, since 1945, by selective breeding.

Animals of the distant past

The spotted hyena, which is still common in East and South Africa, inhabited Britain during a warm period about 150,000 years ago. It probably disappeared during a following cold period, along with other warm-climate species, including the elephant, hippopotamus, macaque monkey and lion.

The giant deer also died out during a cold period, becoming extinct some 13,000 years ago. It had huge, broad-bladed antlers nearly 12 ft (3·6 m) across, and is known chiefly from skeletons found in Irish bogs, which is why it is also called the Irish elk.

Cold-climate animals that died out hundreds of thousands of years ago include the cave bear and woolly rhinoceros. The woolly mammoth, lemming, and arctic fox are among those that disappeared from Britain at the end of the last Ice Age some 10,000 years ago. The woolly mammoth is now extinct, but well-preserved remains are still found deep-frozen in Siberia.

Evidence of old inhabitants

Fossil remains of many animals of the past are found in caves where they were dragged by scavengers such as hyenas and cave bears. They are often mixed with the remains of the scavengers themselves. Some bones are discovered during quarrying or archaeological digs – perhaps those of animals that fell into rock fissures or crevices now full of rocks and debris. Even large animals such as elephants seem to have died in this way.

Some of the larger animals of the past, especially elephants and bison, became trapped in riverside mud as they drank or waded in the water. Today their bones are uncovered in clay pits. Some animals, mammoths for example, were washed downstream in the meltwater from the ice caps, to become buried in riverside sand or gravel banks. Their bones are sometimes revealed when foundations are dug for new buildings in the gravel beds that underlie much of central London and other riverside cities.

Giant deer
Megaceros giganteus

Woolly mammoth
Elephas primigenius

Spotted hyena
Crocuta crocuta

Wild ox
Bos primigenius

Wetland world of the vanished coypu

Marshland and waterside areas are attractive to many mammals because they offer good shelter and varied food. Waterside plants are often luxuriant and include dense masses of tall rushes, reeds and sedges where animals can hide. Clumps of alder growing at the water's edge provide roosting spots for bats in crannies behind fissured bark. On wetlands, most animals are safe from disturbance because few people, and fewer vehicles, venture far over squelchy ground.

One wetland animal that has not survived in the wild is the coypu. It was introduced to Britain from South America about 1929 and reared on fur farms for its pelt – the soft, dense underfur called nutria. Many animals escaped, and colonies became established in the wild, giving Britain its largest rodent, about the size of a short-legged terrier dog.

Most coypus were confined to the wetlands of East Anglia, where they lived in riverbank burrows or in nests of marsh plants and grass. Large burrows – perhaps 6-7 yds (6 m) long – undermined riverbanks, bringing a major flood risk. Coypus ate aquatic plants as well as making nests of them, virtually wiping out some species, and raided farmers' fields for crops. A long campaign to eradicate the coypu began in 1980, and at the end of 1988 the animal was officially declared extinct.

Chinese water deer lie up in extensive reed-beds during the day. They emerge at dawn and dusk to eat crops or lush grass at field edges.

The discharge from drainage pipes includes agricultural chemicals that have seeped through the soil. Fertilisers cause excess growths of algae (plants such as blanket-weed), which use up all the oxygen in the water; other forms of life then die. Pesticides kill large numbers of water insects.

Water voles were often seen swimming by day, leaving a trail of V-shaped ripples. They are now extinct or very rare, because of the degradation of their habitat and predation by mink.

Wetlands have a high population of insects, which provide food for dragonflies by day and bats by night. Typically, Daubenton's bats are out about sunset, pursuing insects very close to the water surface.

Stands of reeds or tall grass close to the water's edge may be the site of a harvest mouse's summer nest. Harvest mice live among tall, stiff-stemmed vegetation and are not confined to farmland.

Footprints that may be seen in the soft mud at the water's edge include the tiny toe marks of a water vole, the pawprints of a mink – or until the late 1980s – the big webbed tracks of a coypu.

The now exterminated coypus normally fed at night but could be seen at dawn or dusk. They ate waterside plants, but also raided nearby farm crops.

The future of Britain's animals

The fortunes of different animals wax and wane, but many are becoming increasingly scarce. Certain uncommon species have been given legal protection (see pp. 297–8), often to bring Britain into line with European law. But while laws may protect individual animals from being shot (for instance), the real cause of decline may be quite different, and not covered by the law at all.

Threats from the environment

For example, the brown hare is not a legally protected species, yet it is becoming rare in many parts of southern Britain. Its decline has much to do with the way that modern farming encourages high densities of sheep and cattle, or the cultivation of huge fields under a single crop. Both are a threat to the hare's food supplies. The law cannot help here because the hare's future depends on the way that farmland is managed.

Many other species are threatened in a similar way. A factor in the decline of the red squirrel may be the clear-felling of conifer woodland, and the consequent loss of its favourite habitat and food supplies. Dormouse populations suffer from the uprooting of hedgerows and the lack of coppicing. Hedgehogs do well on traditionally managed pasture, but increasing economic pressures have caused farmers to plough up fields to grow cereal crops. These are subjected to heavy chemical treatment designed to rid them of pests, but these pests are the very things that hedgehogs need for food. As a result, hedgehogs are now less common than they were. Perhaps the situation will change again as land is taken out of cereal production to reduce agricultural surpluses.

Threats from nature

The water vole is another species that is in decline, apparently as a result of rivers being dredged and lined with concrete banks. However, the continued spread of the mink also seems to threaten water voles, which fall easy prey to it. Granting legal protection to the water vole helps little and its future is uncertain. Similarly, there is little that can be done to help animals that are hard hit by cold or wet summers, such as the reptiles, bats and dormice. A bad summer can seriously affect their breeding performance, and a succession of such summers can decimate populations and even, perhaps, cause small, isolated populations to die out altogether.

Threats from disease

Animals may also fall victim to disease. The 'seal plague' of 1988 caused the deaths of more than 13,000 common seals around European coasts. Pollution of the seas by industrial chemicals was blamed, but there was little evidence to support this. In fact the cause was a virus similar to that which causes distemper in dogs, but its origin is unknown. It is impossible to guard against a completely new disease causing havoc among one wildlife species or another.

Hope for the future

Although many British species are in difficulties, others survive and prosper. Some are multiplying rapidly – the muntjac and the badger among them. Even the vulnerable otter may be staging a comeback in some places, aided by introductions of captive bred animals. The released animals have bred in the wild and their offspring have also bred in their turn.

Other reintroductions have been proposed as a way of re-establishing certain species in areas where they have become extinct. A well-publicised attempt in the early 1980s to reintroduce red squirrels into Regent's Park in London failed, probably because too few animals were available. The habitat was also unsuitable and the presence of grey squirrels and hungry cats did not help. But sand lizards and dormice have been reintroduced to areas from which they had disappeared; and successful reintroductions of pine martens may follow.

Some places to see animals

Wildlife parks, deer parks and rare breed farms offer the most certain opportunities of seeing animals. In the parks, the animals, although captive, wander fairly freely and lead more or less natural lives. They are not as shy as animals in the wild and their normal behaviour can often be observed at length from quite close quarters without their becoming alarmed. Some parks with wild and domestic species, rare breed farms, deer parks (p. 308), nature reserves (p. 311) and forests and open spaces (p. 312) are listed here, as well as some points on the coast where you may see seals (p. 313).

Wildlife parks and rare breed farms

The animals on display and the opening times may be varied without notice. There are usually admission charges. For wildlife parks, only the British species are listed. Many specialise mainly in foreign species.

Acton Scott Working Farm Museum
Shropshire.
3½ miles south of Church Stretton, just off A49. Working horses, longhorn and shorthorn cattle, Shropshire sheep, Tamworth pigs and poultry. Open Apr to Oct 10-5 daily, except Monday. Open Bank Holidays.

Appleby Castle Cumbria.
12 miles north-east from M6, exit 38, by B6260. Sheep, goats, geese, ducks and some waterfowl. Breeds change constantly. Open Easter to end October; daily 10-5, but in October 10-4.

Camperdown Wildlife Centre Dundee.
At Camperdown Park, 4 miles north-west of the centre of Dundee on A923. Red and fallow deer, wolves, reindeer, mouflon, European genet, brown bear, lynx, foxes, wild cats, pine martens, polecats, otters, Shetland ponies, Soay sheep, goats and poultry. Open daily 10-4.30 summer, 10-3.30 winter, except 25-26th Dec and 1-2 Jan.

Chillingham Wild Cattle Park
Chillingham, Northumberland.
4 miles east of Wooler, south off B6348. Wild white Chillingham cattle in park. Open Apr to Oct, 10-12, 2-5 daily except Sun a.m. and Tues.

Cholmondeley Castle Gardens Cheshire.
6 miles north of Whitchurch, just off the A49. Longhorn cattle, many breeds of sheep and pigs; other farm animals, llamas, goats and Shetland ponies. Open Apr to Sept, Wed, Thurs and Sun, and Bank Holidays, 11.30-5.

Cotswold Farm Park Guiting Power, Gloucestershire.
Off B4077, 5 miles west of Stow-on-the-Wold. *Sheep*: Soay, Shetland, North Ronaldsay, Manx Loghtan, Hebridean, Jacob, Castlemilk Moorit, Black Welsh Mountain, Cotswold, Herdwick, Wensleydale, Portland, Norfolk Horn. *Goats*: Bagot, Golden Guernsey, English, Coloured Angora. *Cattle*: Longhorn, Highland, British White, Belted Galloway, White Park, Old Gloucester, Dexter. *Pigs*: Tamworth, Gloucester Old Spot and Iron Age. *Horses and ponies*:

Exmoor, Shetland, Shire. *Rabbits*: Dutch. *Poultry*: Geese, ducks, chickens and turkeys. Open daily Apr to 1 Oct, 10.30-5; Sun and Bank Holidays 10.30-6.

Croxteth Hall and Country Park
Liverpool.
5 miles north-east of Liverpool, in Muirhead Avenue, off A580. Rare cattle, horses, sheep, pigs and poultry. Open daily all year, 11-5.

Dartmoor Wildlife Park and West Country Falconry Devon.
6 miles from Plymouth by A38 then 2 miles on a minor road branching north to Sparkwell. More than 100 species of animals and birds, including tigers, lions, bears, wolves and birds of prey. Also sika, fallow, red, muntjac and Chinese water deer, foxes, grey squirrels, grey seals, adders and grass snakes. Open daily 10-6.

Domestic Fowl Trust Honeybourne Pastures, Honeybourne, Evesham, Worcestershire.
In Honeybourne village, off B4035, 2 miles north of Weston Subedge. *Poultry*: Croad Langshan, Dorking, Marans, Buff Orpington, Old English Pheasant Fowl, Rhode Island Red, Light Sussex, Bronze turkey and almost 150 other species. *Goats*: Bagot and Golden Guernsey. *Pigs*: Gloucester Old Spot. *Sheep*: Balwen, Manx Loghtan, Greyfaced Dartmoor and Hebridean. *Cattle*: Irish Moiled, Shetland, Dexter. More rare breeds of farm animal planned. Open daily 10.30-5.

Easton Farm Park Woodbridge, Suffolk.
Take A12 north from Woodbridge and turn left onto B1116 near Wickham Market; Easton is 2 miles west off B1116. *Horses and ponies*: Suffolk Punch, Shetland. *Pigs*: Gloucester Old Spot. *Cattle*: White Park, Friesian, Longhorn, Red Polls. *Sheep*: Suffolk, Jacob. *Goats*: British Saanen, Pygmy. Pets' paddock with guinea pigs, rabbits, ducks, lambs and peacocks. Closed Mondays except Bank Holidays, but open seven days a week July-Aug, 10.30-6.

Farway Countryside Park Holnest Farm, Farway, near Colyton, Devon.
6 miles south of Honiton, by A375. *Sheep*: Jacob. *Pigs*: Vietnamese Pot-bellied and wild boar. *Deer*: Fallow and red. *Goats*: Pygmy. *Ponies*: Miniature and Shetland. Also poultry. Open Apr to end Sept, 10-5.30 daily.

Highland Wildlife Park Kincraig, Highland.
7 miles south-west of Aviemore on B9152. Reindeer, red and roe deer, Highland cattle, wolves, golden eagles, wild horses, wild cats, foxes, badgers, pine martens, polecats, wild boar, European bison, otters and Soay sheep. Open all year: Nov to end Mar, 10-2; Apr to end June, 10-4; July and Aug, 10-5; Sept and Oct, 10-4 weather permitting. Not open in snow.

Horton Park Children's Farm
Epsom, Surrey.
North-west of Epsom, 1 mile north on B280. Farm animals such as cattle, sheep, pigs, goats and poultry, plus Shetland ponies, rabbits, guinea pigs and reptiles. Lots of young animals for children to see. Open daily, summer 10-6, winter 10-5. Closed Christmas Day and Boxing Day.

Mole Hall Wildlife Park Essex.
South of Saffron Walden, 2½ miles off B1383 in Widdington. Ponies, poultry, otters, Scottish wild cats, wallabies, red squirrels, Shetland ponies, and red, fallow, sika and muntjac deer. Also a butterfly house with tropical and native species. Open daily 10.30-6 or dusk in winter.

Norfolk Wildlife Park
Great Witchingham, Norfolk.
12 miles north-west of Norwich, east of A1067. Grey squirrels, Arctic foxes, polecats, badgers, otters, Patagonian cavies, fallow, red, muntjac and Chinese water deer. Open Apr to Oct daily 10.30-6 or sunset.

Otter Trust Earsham, Suffolk.
1 mile south-west of Bungay, off A143. European and Asian otters, muntjac deer, waterfowl and other wildlife. Open 1 Apr to 31 Oct, daily 10.30-5.30 or sunset if earlier. Other Otter Trust sites are: Otter Trust North Pennines Reserve, Bowes, Barnard Castle, Northumberland and Tamar Otter Sanctuary, Petherwin, near Launceston, Cornwall.

Paradise Park Hayle, Cornwall.
Take A30 south from Hayle; in ¼ mile east onto B3302. Bagot goats, pigs, otters, rabbits, red squirrels and many birds including endangered species of parrot. Open daily 10-5.

Staunton Country Park Havant, Hampshire.
On north side of Havant, west of B2149. *Cattle*: Longhorn crosses. *Sheep*: Hampshire Downs, Dorset Horn, Jacob. *Pigs*: Gloucester Old Spot, Middle White, Vietnamese Pot-bellied, Tamworth Saddleback. *Goats*: Pygmy, Bagot, Angora. *Horses*:

Shire and Fallabella. Also poultry, fallow deer, rabbits, guinea pigs and water buffalo. Open summer 10-5, winter 10-4.

Tilgate Park Nature Centre
Crawley, West Sussex.
2 miles south of Crawley town centre, east of A23 and north of M23 junction 11. White Park cattle, Shire horses, Bagot goats, pigs, Boreray sheep, poultry, small wildlife and rare breeds collection, including red squirrels. Open daily, 10-6 summer, 10-4 winter. Closed winter public holidays.

Wimpole Home Farm Wimpole Hall, Arrington, near Royston, Cambridgeshire.
6 miles north of Royston, just east of A1198. *Cattle*: Irish Moyled, Gloucester, Shetland, Dexter, White Park, Longhorn. *Sheep*: Soay, Hebridean, Leicester Longwool, Portland, Manx Loghtan, Norfolk Horn, Whitefaced Woodland. *Goats*: Bagot, British Alpine, Anglo-Nubian. *Pigs*: Tamworth, Berkshire, Gloucester Old Spot. *Horses*: Suffolk Punch, Shire. *Poultry*: Norfolk Black and Lavender turkeys. Open mid March to 1 Nov, daily 10.30-5 except Mon and Fri, but open daily except Mon in July and Aug. From Nov to Mar, open Sat and Sun 11-4, except Dec 25-27.

Deer Parks
The opening hours may be varied without notice. Where a park adjoins a house, car parks may be open only at house visiting times and cars may not be allowed in parks. There may be admission charges.

Attingham Park Shropshire.
4 miles south-east of Shrewsbury on south side

of A5. Fallow deer to be seen. Owned by the National Trust. Open daily during daylight hours, except Christmas Day.

Blair Drummond Safari and Leisure Park Stirling.

6 miles north-west of Stirling on A84; exit 10 off M9. Fallow, sika and Père David's deer. Open Mar to Oct, daily 10-5.30.

Bradgate Country Park

Newtown Linford, Leicestershire.
5 miles north-west of Leicester by B5327. Red and fallow deer. Open daily to pedestrians. Pay and display car park and visitor centre.

Bushy Park Hampton, Greater London.

West of Kingston upon Thames by A308, opposite Hampton Court Palace. Fallow and red deer. Open daily.

Charlecote Park Warwickshire

5 miles east of Stratford-upon-Avon, off north side of B4086. Fallow and red deer (also Jacob sheep). Open Apr to Oct, Fri to Tues, 12-6. Owned by the National Trust.

Chatsworth Park

Derbyshire.
3 miles north-east of Bakewell, off A619. Fallow and red deer. Open daily from 3rd week in March to end October.

Combe Sydenham Country Park

Somerset.
4 miles south of Watchet on B3188 at Monk-silver. Fallow deer (sometimes Exmoor red deer). Open Easter to Oct, daily 10-5.

Cricket St Thomas Wildlife Park

near Chard, Somerset.
3 miles east of Chard on A30. Fallow, axis and sika deer. Pets corner with guinea pigs, rabbits and rare breeds of sheep, goats and chickens. Open daily 10-6 or dusk in winter.

Culzean Castle and Country Park

South Ayrshire.
12 miles south-west of Ayr off A719. Roe deer roam wild in country park; enclosed deer park within the country park has red deer. Shop and visitor centre. Open 10.30-5.30. Owned by the National Trust for Scotland.

Dunham Massey Park Cheshire.

3 miles south-west of Altrincham. Fallow deer. Open always. Owned by the National Trust.

Dyrham Park near Chippenham, Wilts.

8 miles north of Bath, west off A46, exit 18 off M4. Fallow deer. Open daily 12-5.30 or dusk if earlier. Owned by the National Trust.

Eastnor Castle Estate Herefordshire.

3 miles south-east of Ledbury by A438. Red deer. Deer park open from Easter to end Sept, dawn to dusk.

Glengoulandie Deer Park Perthshire and Kinross.

North-west of Aberfeldy on B846 between Weem and Tummel Bridge. Red and fallow deer. Also Highland cattle, Soay sheep. Open Easter to end Sept, from 9 to two hours before dusk.

Hampton Court Home Park Greater London.

By Hampton Court Palace; entrance also at Kingston Bridge. Fallow deer. Open daily.

Holker Hall Cark-in-Cartmel, Cumbria.

17 miles south-west of Kendal via A590 from Levens, then B5277. Fallow deer. Open 1 Apr to 31 Oct, daily except Sat, 10-6.

Holkham Park Wells, Norfolk.

2 miles west of Wells-next-the-Sea on A149. Fallow deer. Open daily, but no vehicular access when house is closed.

Hopetoun House

Queensferry, Edinburgh.
2 miles west of Forth Bridge off A904. Fallow and red deer, also St Kilda sheep. Open Easter to Sept, daily 10-5.30.

Howletts Wild Animal Park

Bekesbourne, Kent.
3 miles south-east of Canterbury, just off A2. Axis deer and other foreign species of deer. Open 10-5 daily, except Christmas Day.

Knebworth Park Hertfordshire.

1 mile south of Stevenage, south-west of A1(M) exit 7; 12 miles north of M25. 250 acres of grassland with more than 100 red and sika deer. Open Easter, then Whitsun to Sept daily, 11-5.30.

Knole Park Sevenoaks, Kent.

At south end of Sevenoaks, just east of A225. Fallow and sika deer. Owned by National Trust, but park open daily to pedestrians by courtesy of Lord Sackville. Parking at Knole House, open from 1 Apr to end Oct, Wed to Sat, 12-4; Sun, Good Friday and Bank Holiday Mondays, 11-5.

Levens Hall Park near Kendal, Cumbria.

5 miles south of Kendal off A6. Fallow deer and

Bagot goats. Public footpath open daily (no cars or parking allowed in the park). The Hall is open Apr to mid-Oct, Sun to Thurs.

London parks with fallow deer enclosures. Clissold Park, Stoke Newington N16; Golders Hill Park, Golders Green NW3; Greenwich Park SE10; Maryon Wilson Park, Charlton SE7.

Lough Key Forest Park Boyle, Co. Roscommon, Republic of Ireland. 2 miles east of Boyle on Carrick Road (N4). Fallow deer enclosure. Open daily, all year.

Lyme Park Disley, Cheshire. 6½ miles south-east of Stockport on south side of A6. Red and fallow deer. Open daily Apr to Oct, 8-8.30; Nov to Mar 8-6. National Trust owns it.

Margam Country Park Neath and Port Talbot. 2 miles south-east of Port Talbot off A48 between Margam and Pyle. Junction 38, M4. Fallow, red, Père David, muntjac, barasingha and Chinese water deer; also wild boar, foxes, squirrels, Glamorgan cattle, Iron Age pigs, ponies, sheep. Open Apr to Sept, daily 10-8, last entry 5pm. Oct to Mar, Wed. and Sun from 10 to one hour before sunset.

Marwell Zoological Park Hampshire. 6 miles south-east of Winchester on B2177. Hog, muntjac and axis. Open daily 10-6 summer, 10-dusk winter. Closed Christmas Day.

Normanby Hall Country Park North Lincolnshire. 5 miles north of Scunthorpe by B1430. Red and fallow deer. Open daily.

Parkanaur Forest Park Dungannon, Co. Tyrone, Northern Ireland. On south-west side of Lough Neagh. White fallow deer. Open daily 8 to sunset. There is also a small enclosure for sika deer at Seskinore Forest, 7 miles south-east of Omagh.

Petworth Park Petworth, West Sussex. 5½ miles east of Midhurst on A272. Owned by National Trust. Fallow deer. No vehicles allowed in park. Open daily, 8 to sunset.

Phoenix Park Dublin, Republic of Ireland. On north-west side of the city, off Conyngham Road. Largest enclosed deer park in Europe. 600 fallow deer. Open Mar to end Oct, daily 10-5. Weekends only in other months.

Port Lympne Wild Animal Park Kent. 8 miles south-east of Ashford, just south of A20; exit 11 off M20; near Hythe. Axis, barasingha, elds and samba deer. Open daily 10-5 except Christmas Day.

Randalstown Forest Randalstown, Co. Antrim, Northern Ireland. 2 miles south-west of Randalstown on minor road to Staffordstown. Fallow deer. Open daily.

Richmond Park Richmond-upon-Thames, London. Enter from Richmond, Sheen, Roehampton, Ham and off A308. Fallow and red deer. Open daily.

Sewerby Hall and Gardens Bridlington, East Yorkshire. 1 mile north-east of Bridlington off B1255. Sika deer in paddocks. Open daily.

Spetchley Park Worcestershire. 3½ miles east of Worcester via A422. Fallow and red deer. Open Apr to Sept, weekdays except Mon and Sat, 11-5; Sun 2-5, Bank Holidays 11-5.

Stonor Park Oxfordshire. 5 miles north of Henley-on-Thames via A4130 and B480. Fallow deer. Open 2-5.30, on Sun in Apr to Sept; in addition, Wed in July and Aug, plus Bank Holiday Mondays.

Studley Royal Park North Yorkshire. By Fountains Abbey, on a minor road south from B6265, 4 miles south-west of Ripon. Red, sika and fallow deer. Open daily in daylight hours. Owned by the National Trust.

Tatton Park Knutsford, Cheshire. 3½ miles north of Knutsford. Fallow and red deer. Open Apr to end Oct, 10-6 daily; 2 Nov to Mar, Tues to Sun, 11-4. National Trust owns it.

Weston Park Shifnal, Shropshire. On A5, 7 miles west of Gailey (A5/M6 junction). Fallow deer. Open Bank Holidays, Easter to Sept.; May to June weekends only; July daily except Mon and Fri; Aug daily.

Whipsnade Wild Animal Park Bedfordshire. Signed from M25 (junction 21) and M1 (junctions 9 and 12). Fallow, muntjac, sika and Chinese water deer. Also foreign species including Axis and Père David's deer. Open Easter to mid-Sept, daily 10-6, Sun 10-7.

Woburn Abbey and Deer Park Woburn, Bedfordshire.

6 miles north of Leighton Buzzard by A4012; exit 13 from M1. Fallow, Marchunian, sika, red, muntjac, rusa, axis, Père David's and Chinese water deer. Plus black varieties of grey squirrel. Open Apr to end Sept, daily 10-4.45; Oct weekends only; Nov and Dec closed; Jan to Mar weekends 10.30-3.45.

Wollaton Hall Nottinghamshire.
On A609 on western edge of Nottingham. Red and fallow deer. Park open daily, dawn to dusk. Closed Dec 25-26 and Jan 1.

Nature reserves

There are many nature reserves where wild animals may be seen, with patience and good fortune. Nature reserves are primarily intended for the conservation and study of wildlife, but visitors may be allowed. Always keep to marked footpaths and do not enter fenced-off areas. The reserves listed are open to visitors, but parties should make arrangements with the wardens. Some reserves ask an entry fee.

National Nature Reserves (marked NNR) are administered by English Nature, Scottish Natural Heritage or the Countryside Council for Wales. National Trust properties are marked NT or NTS (National Trust for Scotland). Maritime Nature Reserves are marked MNR. Other reserves are administered either by local government authority or by naturalists' trusts.

Ainsdale Sand Dunes (NNR) Lancashire.
3 miles north of Formby off A565 Southport road. Sand lizards, common lizards, natterjack toads, red squirrels, foxes, rabbits, butterflies, moths and dragonflies. Also Herdwick sheep

(Oct-March only). No vehicles or horses allowed; public access by marked path only or by permit from English Nature.

Ashridge Estate (NT) Herts/Bucks.
4 miles north of Berkhamsted on B4506. Fallow and muntjac deer.

Beinn Eighe (NNR) Highland.
1 mile south-west of Kinlochewe on A896. Pine martens, wild cats, red and roe deer.

Bridestones Moor (NT) North Yorkshire.
6 miles north-east of Pickering by A170, then minor road north from Thornton-le-Dale. Access to the moor is by Dalby Forest Drive, for which there is a small toll. Roe deer, adders, common lizards, slow-worms, foxes and badgers may be seen.

Brodick Country Park and Goat Fell
(NTS) Isle of Arran, North Ayrshire.
2 miles north of Brodick. Red deer, red squirrels and badgers can be seen.

Brownsea Island (NT) Dorset.
Reached by boat from Poole, Bournemouth and Sandbanks. Dorset Wildlife Trust gives guided tours of the reserve. Red squirrels and sika deer. Open April to first week October.

Cairngorms (NNR) Highland.
10 miles south-east of Aviemore. Red and roe deer, reindeer, wild cats, red squirrels, mountain hares, badgers, otters and pine martens.

Cwm Idwal (NNR) Gwynedd/Conwy.
5 miles west of Capel Curig by A5. Feral goats.

Ebbor Gorge (NNR) Somerset.
3 miles north-west of Wells, east off A371. Badgers, foxes and roe deer.

Garbutt Wood Sutton Bank, North Yorkshire.
4 miles east of Thirsk on A170, close to North Yorkshire Moors National Park visitor centre. Managed by Yorkshire Wildlife Trust. Badgers, foxes and roe deer.

Gibraltar Point Lincolnshire.
3 miles south of Skegness by minor road. Rabbits; also common seals on sandbanks.

Glen Coe and Dalness (NTS) Highland.
Access from A82 east of Glencoe. Red deer and wild cats. There is a visitor centre at Glen Coe.

Grey Mare's Tail
(NTS) Annandale, Dumfries and Galloway. Spectacular 200 ft. waterfall 10 miles north-east of Moffat off A708. Feral goats may be seen. There is a visitor centre with a remote CCTV link which allows visitors to observe peregrine falcons nesting.

Hatfield Forest (NT) Essex.
3 miles east of Bishop's Stortford on A120. Fallow and muntjac deer, badgers, weasels, foxes bats.

Inkpen Common Newbury.
3½ miles east of Hungerford. Adders, common lizards. Managed by Berks, Bucks and Oxon Wildlife Trust.

Inverpolly (NNR) Highland.
12 miles north of Ullapool, to west of A835. Red deer, seals, weasels, stoats and otters.

311

Useful addresses

BAT CONSERVATION TRUST *Cloisters House, 8 Battersea Park Road, London SW8 4BG. www.bats.org.uk* The only UK organisation solely devoted to the conservation of bats and their habitat. Helps bats through practical conservation projects and research, by encouraging appreciation and enjoyment of bats and advising people who find bats in their property.

BRITISH DEER SOCIETY *Burgate Manor, Fordingbridge, Hampshire SP6 1EF. www.bds.org.uk* Promotes the study and welfare of wild deer, their management, conservation and protection through education, research and training courses.

BRITISH HEDGEHOG PRESERVATION SOCIETY *Knowbury House, Knowbury, Ludlow, Shropshire SY8 3LQ. www.argyll.demon.co.uk/bhps.html* Gives advice to the public on the care of injured, sick and orphaned hedgehogs and can supply names of carers in local areas.

BRITISH HERPETOLOGICAL SOCIETY *c/o Zoological Society of London, Regent's Park, London NW1 4RY.* A society for the scientific study of reptiles and amphibians both in captivity and in the wild.

BRITISH NATURALISTS' ASSOCIATION *P.O. Box 5682, Corby, Northamptonshire NN17 2RT.* Supports schemes and legislation for the protection of wildlife and the promotion and maintenance of national parks and nature reserves. Supplements the activities of local and regional bodies.

Produces *Countryside* magazine six times a year and is associated with the Blake Shield BNA Trust natural history competition for young people aged 7-18.

BTCV *36 St Mary's Street, Wallingford, Oxfordshire OX10 0EU. www.btcv.org* Involves volunteers of all ages in practical conservation work in local groups nationwide. Offers an extensive programme of UK and international conservation holidays.

BTCV SCOTLAND *Balallan House, 24 Allan Park, Stirling FK8 2QG. www.bctv.org* Organises volunteer groups to undertake practical, environmental conservation throughout the whole of Scotland.

COUNTRYSIDE COUNCIL FOR WALES *Plas Penrhos, Ffordd Penrhos, Bangor, Gwynedd LL57 2LQ. www.ccw.gov.uk* Promotes nature conservation in Wales; advises on wildlife conservation; selects, establishes and manages National Nature Reserves. Issues licences for exemption from the provisions of the Wildlife and Countryside Act 1981. Publishes leaflets and books.

ENGLISH NATURE *Northminster House, Peterborough PE1 1UA. www.english-nature.org.uk* Promotes nature conservation in England; advises on the conservation of wildlife; and selects, establishes and manages National Nature Reserves. Issues licences for exemption from the provisions of the Wildlife and Countryside Act 1981. Publishes leaflets and books.

FAUNA AND FLORA INTERNATIONAL *Great Eastern House, Tenison Road, Cambridge CB1 2IT. www.fauna-flora.org* Acts to conserve threatened species of wild animals and plants worldwide, choosing solutions that are sustainable, based on sound science and taking account of human needs.

FIELD STUDIES COUNCIL *Head Office: Preston Montford, Montford Bridge, Shrewsbury SY4 1HW. www.field-studies-council.org* Encourages field work and research in all branches of knowledge concerned with the outdoors. There are many courses at the council's residential centres.

MAMMAL SOCIETY *15 Cloisters Business Centre, 8 Battersea Park Road, London SW8 4BG. www.mammal.org.uk* This society works to protect and promote interest in all British mammals and halt the decline of threatened species. It publishes a newsletter, a journal (*Mammal Review*), notes, booklets and fact sheets. The society has a network of local groups, and organises conferences and regional meetings. The society's youth group, known as Mammalaction, is open to young people under the age of 18; it also organises activities and publishes a newsletter.

NATIONAL TRUST *36 Queen Anne's Gate, London SW1H 9AS. www.nationaltrust.org.uk* Registered charity, founded in 1895 to preserve places of historic interest or natural beauty for the nation to enjoy. Now owns more than 612,000 acres of countryside, almost 600 miles of coastline, more than 200 historic houses and gardens, and 49 industrial monuments and mills. Publishes a list of properties in England, Wales and Northern Ireland.

AN TAISCE (NATIONAL TRUST FOR IRELAND) *The Tailors' Hall, Back Lane, Dublin 8. www.antaisce.org* This state charity and voluntary organisation conserves and protects the best of Ireland's natural environment and man-made heritage.

NATIONAL TRUST FOR SCOTLAND *28 Charlotte Square, Edinburgh EH2 4ET. www.nts.org.uk* Publishes a list of its properties open to the public. These include large tracts of land which are of interest to naturalists.

RARE BREEDS SURVIVAL TRUST *National Agricultural Centre, Stoneleigh Park, Kenilworth, Warwickshire CV8 2LG. www.rare-breeds.com* Promotes the conservation of rare breeds of British farm animals.

ROYAL SOCIETY FOR THE PREVENTION OF CRUELTY TO ANIMALS (RSPCA) *The Causeway, Horsham, West Sussex RH12 1HG. www.rspca.co.uk* Exists to promote kindness to animals and to prevent and investigate cruelty to and persecution of them; advises on the care of sick animals. It also runs wildlife hospitals and veterinary centres, rehabilitating injured wildlife and rehousing abandoned or badly treated domestic animals.

SCOTTISH FIELD STUDIES ASSOCIATION *Kindrogen Field Centre, Enochdu, Blairgowrie, Perthshire and Kinross PH10 7PG. www.kindrogen.com* An educational body giving teachers, students and amateurs opportunities to increase their knowledge of the countryside.

SCOTTISH NATURAL HERITAGE *12 Hope Terrace, Edinburgh EH9 2AS. www.snh.org.uk* Promotes nature conservation in Scotland; establishes and manages National Nature Reserves. Advises on policies and supports projects that promote sustainable use; encourages responsible enjoyment of Scotland's wildlife, habitats and landscape.

SCOTTISH WILDLIFE TRUST *Cramond House, Kirk Cramond, Cramond Glebe Road, Edinburgh EH4 6NS. www.swt.org.uk* Organisation concerned with the conservation of all types of wildlife and their habitats in Scotland. It establishes and manages wildlife reserves, conducts surveys and advises on wildlife management and planning.

ST TIGGYWINKLES *Aston Road, Haddenham, Aylesbury, Bucks. HP17 8AF. www.sttiggywinkles.org.uk* Wildlife hospital trust which cares for sick and injured wild animals and birds, with the aim of rehabilitating them to the wild. The public can become members of the trust, and there is an under-12 group called 'The Tiggies'. An interpretation education centre for the public has been opened.

THE WILDLIFE TRUSTS *The Kiln, Waterside, Mather Road, Newark, Nottinghamshire NG24 1WT. www.wildlifetrust.org.uk* This is the co-ordinating body of all local wildlife trusts which own and manage nature reserves, and it is also the parent body of the junior wing, Wildlife Watch.

WORLD WIDE FUND FOR NATURE (WWF) *Panda House, Weyside Park, Godalming, Surrey GU7 1XR. www.wwf-uk.org* WWF takes action to protect the environment for people and nature, through protecting species and spaces. Influences attitudes and behaviour through education, field work, advocacy and partnerships.

INDEX

Acknowledgments

Artwork in *Animals of Britain* was supplied by the following artists:

3, 8–25 Sarah Fox-Davies · 26–27 David Nockels · 28–29 Eric Robson · 30 Phil Weare · 31 David Nockels · 32–33 Jim Russell · 34–39 Peter Barrett · 40–43 Eric Robson · 44–73 Peter Barrett · 74–75 Libby Turner · 76–103 John Francis · 104–7 Eric Robson · 108–19 Peter Barrett · 110–21 Brian Delf · 122–3 Sarah Fox-Davies · 124–31 Gill Tomblin · 132–41 Sarah Fox-Davies · 142–3 Gill Tomblin · 144–51 Sarah Fox-Davies · 152–9 Brian Delf · 160–3 Sarah Fox-Davies · 164–5 Peter Barrett · 166–7 Dick Bonson · 168–9 H. Jacob and Eric Robson · 170–3 Jim Channell · 174–5 Kevin Dean · 176–9 Robert Morton · 180–1 Eric Robson · 182–205 Gill Tomblin · 206–17 Philip Weare · 218–23 Rosalind Hewitt · 224–9 Phil Weare · 230–1 Eric Robson · 232–41 Gill Tomblin · 242–3 Rosalind Hewitt · 244–51 David Nockels · 252–5 Tim Hayward · 256–69 David Nockels · 270–3 Tim Hayward · 274–7 Eric Robson · 278–81 Tim Hayward · 282–3 Robert Morton · 284–7 David Nockels · 288–91 Libby Turner · 292–3 Eric Robson · 294–5 Jim Russell · 306–7 Eric Robson

The publishers wish to thank the following for their help in providing information and reference for artwork: Nature Conservancy Council · Forestry Commission · Ieuan Jones and family, Llanddewi Brefi, Tregaron, Dyfed, Wales · Mrs M. Rosenberg, British Angora Goat Society, Iddesleigh, Devon · S. & G. Fowler, The English Goat Breeders' Association, Arkholme, Carnforth, Lancashire · Domestic Fowl Trust, Stratford-upon-Avon, Warwickshire.

Photographs in *Animals of Britain* were supplied by the following photographers and agencies. Names of agencies are in capital letters. The following abbreviations are used:
KRD – King, Read and Doré.
NHPA – Natural History Photographic Agency.
NSP – Natural Science Photographs.
OSF – Oxford Scientific Films.

35 BRUCE COLEMAN/J. Burton · 41 J. P. Ferrero/ARDEA, LONDON · 43 NHPA/S. J. Harris · 45 NATURE PHOTOGRAPHERS/O. Newman · 51 WILDLIFE SERVICES/M. Leach · 59 NHPA/R. Balharry · 61 BRUCE COLEMAN/J. Burton · 63 M. Wilding/SURVIVAL ANGLIA/OSF · 65 E. A. Read/KRD · 67 BRUCE COLEMAN/J. Burton · 69 NATURE PHOTOGRAPHERS/R. Tidman · 71 R. Redfern/OSF · 75 J. Daniels/ARDEA, LONDON · 83 E. & D. Hosking · 87 BRUCE COLEMAN/G. Langsbury · 91 I. Beames/ARDEA, LONDON · 93 Heather Angel · 95 BRUCE COLEMAN/H. Reinhard · 99 NHPA/M. Clark · 101 FRANK LANE/B. Lawrence · 102 E. & D. Hosking · 103 FRANK LANE/B. Lawrence · 105 BRUCE COLEMAN/E. Breeze-Jones · 107 J. Gooders/ARDEA, LONDON · 109 Dr Yalden · 111 BRUCE COLEMAN/H. Reinhard · 117 BRUCE COLEMAN/J. Burton · 125–7 B. E. Watts/OSF · 129 BRUCE COLEMAN/J. Burton · 131 E. & D. Hosking · 133 I. Beames/ARDEA, LONDON · 135 E. & A. Bomford/SURVIVAL ANGLIA/OSF · 137 G. I. Bernard/OSF · 139 E. & A. Bomford/SURVIVAL ANGLIA/OSF · 141 NATURE PHOTOGRAPHERS/O. Newman · 145 BRUCE COLEMAN/K. Taylor · 147 E. & D. Hosking · 149 E. & A. Bomford/ARDEA, LONDON · 151 AQUILA/G. F. Date · 153 NSP/D. MacGaskill · 157 FRANK LANE/B. Lawrence · 161 AQUILA/J. Robinson · 163 BRUCE COLEMAN/H. Reinhard · 165 WILDLIFE SERVICES/M. Leach · 171 REX FEATURES/S. Meyers · 173 NSP/D. MacGaskill · 177 BRUCE COLEMAN/J. Burton · 185 NATURE PHOTOGRAPHERS/S. C. Bisserot · 187 P. Morris/ARDEA, LONDON · 191 NATURE PHOTOGRAPHERS/S. C. Bisserot · 192 Pat Morris · 193 Sdeuard Bisserot · 195 Brian Hawkes · 197 WILDLIFE SERVICES/M. Leach · 198 Pat Morris · 199 J. Mason/ARDEA, LONDON · 200 J. Mihok/SURVIVAL ANGLIA/OSF · 201 BRUCE COLEMAN/H. Reinhard · 202 J. Mason/ARDEA, LONDON · 203 S. Roberts/ARDEA, LONDON · 207 NHPA/S. Dalton · 209 NATURE PHOTOGRAPHERS/P. R. Sterry · 211 NATURE PHOTOGRAPHERS/O. Newman · 213 BRUCE COLEMAN/H. Reinhard · 215 G. I. Bernard/OSF · 219 R. J. G. Blewitt/ARDEA, LONDON · 221 NHPA/S. Dalton · 223 BRUCE COLEMAN/G. Langsbury · 225 E. & D. Hosking · 229 NHPA/S. Dalton · 229 M. Linley/SURVIVAL ANGLIA/OSF · 246 Anthony Blake · 247 AQUILA/M. C. Wilkes · 248 Lawrence Alderson · 249 John Sims · 250 NSP/J. A. Grant · 251 Alan Beaumont · 256 Wensleydale Longwool Sheep Breeders' Assn · 257 NATURE PHOTOGRAPHERS/ A. K. Davies · 258 WILDLIFE INTERNATIONAL/M. Beckwith · 259 P. Morris/ARDEA, LONDON · 260 Lawrence Alderson · 261 BRUCE COLEMAN/E. Crichton · 270 NATURE PHOTOGRAPHERS/C. Mylne · 271 BRUCE COLEMAN/J. Burton · 274 G. I. Bernard/OSF · 275 NATURE PHOTOGRAPHERS/S. C. Bisserot · 278 BRUCE COLEMAN/J. Burton · 279 Robert Estall · 282 S. Gooders/ARDEA, LONDON · 283–8 J. P. Ferrero/ARDEA, LONDON · 289 Lawrence Alderson

The publishers also acknowledge their indebtedness to the following books and journals, consulted for reference:

The Handbook of British Mammals Ed. G. N. Corbet and H. N. Southern (Blackwell) · *The British Amphibians and Reptiles* M. Smith (Collins) · *The Observers's Book of Farm Animals* L. Alderson (Frederick Warne) · *The Field Guide A Farmland Companion* J. Woodward, P. Luff (Blandford) · *The Living Countryside* weekly nature guide (Orbis) · *The Chance to Survive* L. Alderson (Cameron & Tayleur with David & Charles) · *The Newfoundland* Ed. Carol Cooper (The Newfoundland Club) · *Hedgehogs* P. Morris (Whittet Books) · *The Mammals of Britain and Europe* G. Corbet, D. Ovenden (Collins) · *The Observer's Book of Horses and Ponies* R. S. Summerhays (Frederick Warne) · *The Rabbit* H. V. Thompson, A. N. Worden (Collins) · *The Guinness Book of Mammals* J. A. Burton (Guinness Superlatives) · *Animal Tracks and Signs* M. Bouchner (Octopus) · *Animal Tracks and Signs* P. Bang, P. Dahlstrom (Collins) · *Tracks and Signs of British Animals* A. Leutscher (Cleave-Hume) · *Mammal Watching* M. Clark (Severn House) · *Natural History of the British Isles* Ed. P. Morris (Country Life) · *Wildlife of Scotland* F. G. T. Holliday (Macmillan) · *Wildlife Through the Camera* P. Bolt (BBC) · *The Exmoor Pony* V. G. Speed (Countrywide Livestock) · *European Breeds of Cattle* (Vol 1) M. R. French (F. A. O.) · *Breeds of Goats* H. E. Jeffery (British Goat Soc.) · *British Beef Cattle* (Meat & Livestock Comm.) · *British Sheep* (National Sheep Assoc.) · *The Complete Book of Raising Livestock and Poultry* Ed. Katie Thear, A. Fraser (Martin Dunitz) · *The Old English Game Fowl* H. Atkinson (Saiga) · *The Wild Rabbit* D. Cowan (Blandford) · *British Cattle* (National Cattle Breeders' Assoc.) · *The Red Fox* H. G. Lloyd (Batsford) · *The World of the Red Fox* L. L. Rue (J. B. Lippincott Co.) · *Fox family: four seasons of animal life* Taketazu (Weatherhill) · *Cattle of the World* Friend and Bishop (Blandford) · *Mammals of the World* E. P. Walker (John Hopkins) · *The Badger* E. Dudley (F. Muller) · *My Wilderness Wild Cats* M. Tomkies (Macdonald) · *Wild Goats of Britain and Ireland* G. Whitehead (David & Charles) · *The Harvest Mouse* S. Harris (Blandford) · *The Badger* E. Neal (Collins) · *Badgers* E. Neal (Blandford) · *Squirrels* M. Shorten (Collins) · *Squirrels in Britain* Laidler (David & Charles) · *The Red Squirrel* Tittensor (Blandford) · *Otters in Britain* Laidler (David & Charles) · *Watch for the Otter* E. Hurrell (Country Life) · *The Greater Horseshoe Bat* R. Ransome (Blandford) · *The Cattle of Britain* F. H. Garner (Longmans, Green & Co.) · *The Bull* A. Fraser (Osprey) · *Modern Breeds of Livestock* H. M. Briggs (Macmillan) · *An Illustrated History of Belted Cattle* Lord David Stuart (Scottish Academic Press) · *Natural History of Britain and Ireland* H. Angel (M. Joseph) · *Sheep of the World* K. Ponting (Blandford) · *Island Survivors* Ed. P. A. Jewell (Athlone) · *British Sheep Breeds* (British Wool Marketing Board) · *Mountain Sheep* V. Geist (Univ. of Chicago) · *Practical Goat Keeping* J. & J. Halliday (Ward Lock) · *Keeping Domestic Geese* B. Soames (Blandford) · *British Poultry Standards* Ed. C. G. May (Butterworths) · *Your Sheepdog and its Training* Longton & Hart (Alan Exley) · *The Sheep Dog, its Work and Training* Longton & Hart (David & Charles) · *The Shire Horse* K. Chivers (J. A. Allen) · *Native Ponies of the British Isles* S. Hulme (Saiga) · *Pig Husbandry* J. Luscombe (Farming Press) · *Book of the Pig* S. Hulme (Saiga) · *The Deer of Great Britain* G. K. Whitehead (Routledge and Kegan Paul) · *Deer of the World* G. K. Whitehead (Constable) · *Wild Deer in Britain* R. A. Harris, K. R. Duff (David & Charles) · *Following the Roe* F. Holmes (Bartholomew) · *Red Deer*. T. Clutton-Brock, S. Albon, F. Guinness (Edinburgh U.P.) · *Fallow Deer* D. & N. Chapman (Terence Dalton) · *The Fallow Deer* N. G., D. I. Chapman (HMSO) · *Deer of East Anglia* D. N. Chapman (Pawsy) · *Roe Deer* F. J. Taylor Page (Sunday Times) · *The Red Deer* B. Staines (Blandford) · *Sika Deer* M. Horwood, E. Masters (British Deer Soc.) · *Roe Deer* P. Delap (British Deer Soc.) · *Deer in the New Forest* J. Jackson (Moonraker) · *The Deer and the Tiger* G. Schaller (Univ. of Chicago) · *The Wild Life of India* E. P. Gee (Collins) · *Rehuild* F. Kurt (BLV Munich) · *Das Reh* F. von Oehsen (Landbuch) · *Field Guide to British Deer* F. J. Taylor Page (Blackwell) · *Mammals – Their Latin Names Explained* A. F. Gotch (Blandford) · *Das Tier Sammelband* (Ernst Klett)

Typesetting: VANTAGE PHOTOSETTING CO. LTD, EASTLEIGH
Separations: MULLIS MORGAN LTD, LONDON
Printer/Binder: MILANOSTAMPA, ITALY

400-078-01